Tourism as a Resource-based Industry

FSC
www.fsc.org
MIX
Paper from
responsible sources
FSC® C022174

Sondre Svalastog (1940–2012)

Tourism as a Resource-based Industry

Based on the work of Sondre Svalastog

Edited by

Anna Lydia Svalastog,

Dieter K. Müller

and

Ian Jenkins

CABI

CABI is a trading name of CAB International

CABI
Nosworthy Way
Wallingford
Oxfordshire OX10 8DE
UK

Tel: +44 (0)1491 832111
E-mail: info@cabi.org
Website: www.cabi.org

CABI
WeWork
One Lincoln St
24th Floor
Boston, MA 02111
USA

Tel: +1 (617)682-9015
E-mail: cabi-nao@cabi.org

A catalogue record for this book is available from the British Library, London, UK.

Library of Congress Cataloging-in-Publication Data

Names: Svalastog, Anna Lydia, editor. | Müller, Dieter K, editor. | Jenkins, Ian, editor.
Title: Tourism as a resource-based industry / Edited by Anna Lydia Svalastog, Dieter K Müller & Ian Jenkins.
Description: Oxfordshire ; Boston, MA : CABI, [2022] | Includes bibliographical references and index. | Summary: "Based on Norwegian scholar Sondre Svalastog's work, this book promotes research-based knowledge for the tourism industry, considering social, cultural and natural resources, as well as the importance of local conditions. It provides a wide selection of case studies covering how to maximise potential for both host and tourist's benefit"-- Provided by publisher.
Identifiers: LCCN 2021054080 (print) | LCCN 2021054081 (ebook) | ISBN 9781800621466 (hardback) | ISBN 9781800621473 (ebook) | ISBN 9781800621480 (epub)
Subjects: LCSH: Tourism--Norway--Planning. | Tourism--Planning. | Tourism--Norway--Management. | Tourism--Management.
Classification: LCC G155.N6 T68 2022 (print) | LCC G155.N6 (ebook) | DDC 338.4/791481--dc23/eng20220228
LC record available at https://lccn.loc.gov/2021054080
LC ebook record available at https://lccn.loc.gov/2021054081

ISBN-13: 9781800621466 (hardback)
9781800621473 (ePDF)
9781800621480 (ePub)

DOI: 10.1079/9781800621480.0000

Commissioning Editor: Claire Parfitt
Editorial Assistant: Lauren Davies
Production Editor: Marta Patiño

Typeset by SPi, Pondicherry, India
Printed and bound in the UK by Severn, Gloucester

Contents

Contributors

Øystein Aas is Professor in Nature-based Tourism and Natural Resource Management at the Norwegian University of Life Sciences. His research interests cover a range of fields from human dimensions of natural resource management, ecotourism and nature-based tourism via impact assessment to social acceptance of renewable energy. He has a certain interest in consumptive and non-consumptive human–wildlife interactions, specifically on recreational fisheries and Atlantic salmon conservation, and he has edited and published two peer-reviewed books on these topics. He has been actively involved in a range of inter- and transdisciplinary research projects funded by the EU's H2020 programme and the Norwegian Research Council, as well as taking on editorial duties in scientific journals such as *Human Dimensions of Wildlife* and *Energy Research and Social Sciences*.

Lars Aronsson is Professor of Human Geography, Karlstad University, with research interests in tourism; tourism geography; the concepts of place, time and space; local and regional development and planning; sustainable development and the mobile society; cultural economy; cultural planning; creativity and place; place branding; communication; and representation and experience of places.

Sjur Baardsen is currently Professor, Faculty of Environmental Sciences and Natural Resource Management, Norwegian University of Life Sciences. His research interests cover a range of fields from nature-based tourism to econometric modelling of markets and firms. He worked on rural development and forestry issues for the DG RTD of the European Commission (2000–2004) and chaired the former COST Domain Committee for Forests, their Products and Services (2006–2014). Sjur acted as Rector for the University (2019–2021) and as Dean (2014–2019). In his earlier career he had short appointments in a forest-owner association and in the Norwegian Ministry of Agriculture. Professor Baardsen has performed several roles, including President, for the former European Centre for Nature Conservation (2006–2017). He has been the main supervisor for numerous PhD and Master's theses.

Børge Dahle is a retired senior lecturer in the field of leisure at the Norwegian School of Sport Science. His work is rooted in a lifelong engagement with the environmental movement, as expressed with his friend and colleague, the Norwegian professor of philosophy and founder of deep ecology, Arne Næss. The use of the concept *friluftsliv* ('free, air, life') in the area of leisure studies stems from Dahle's work. 'Free, air, life' is rooted in Norwegian culture and tells us that the key to our wellbeing lies not just outdoors but also in a close connection to nature. In English, his perspective is presented in the anthology *Nature First: Outdoor Life the* Friluftsliv *Way* (Natural Heritage Books, Toronto).

Ian Jenkins was Associate Professor at the Department of Geography and Tourism, University of Iceland, and is currently a visiting associate professor there. He has also worked in Wales and Switzerland as a researcher and academic at a number of universities. Published research includes in the areas of festivals, sustainable tourism, literary tourism, adventure tourism, together with risk and safety management. He met Sondre Svalastog at a tourism conference in Lillehammer and spent a wonderful week with him touring the beautiful sites of southern Norway, a memory he has always cherished.

Marko Koščak is an Associate Professor at the University of Maribor, Faculty of Tourism, Brežice, Slovenia. He studied in Ljubljana, Birmingham and Vienna, and in 1999 completed his PhD on 'Transformation of Rural Areas Along the Slovene–Croatian Border' at the University of Ljubljana. His academic interests are in the fields of sustainable and community tourism, geography of tourism, and destination management. On these he lectures. He met Sondre in 1992 in Lillehammer, together with Professor Travis, two years prior to the Olympic Games, and stayed in touch, discussing the issues of sustainable tourism development in Norway, Slovenia and elsewhere.

Brian McNeil studied at the University of Cambridge, where he took his doctorate in Syriac patrology, in 1978, and a licentiate in theology from the Angelicum University in Rome. He became a Catholic priest. He has lived and worked in Italy, Norway (where he became a friend of the Svalastog family) and Germany. He has translated more than 70 books and many articles into English.

Dieter K. Müller is Professor of Human Geography, Umeå University, with research interests in tourism and regional development, and mobility and tourism in peripheral areas. His research specifically includes almost all aspects of second homes and second home-related mobility; Sami tourism; nature-based tourism; tourism labour markets; and regional development and rural change, particularly in northern peripheries and polar areas. From 2008 to 2020, he was a member and Chair of the International Geographical Union's Commission on Tourism, Leisure and Global Change, and in 2018 the Specialty Group for Recreation, Tourism and Sports of the American Association of Geographers presented him the Roy Wolfe Award for outstanding contribution in the field.

Stian Stensland is an associate professor in nature-based tourism at the Norwegian University of Life Sciences. His research is in the fields of natural resources management and nature-based tourism and the linkage between these two as a prerequisite for sustainable development. His core research has been in fishing tourism, nature-based tourism firms and the human dimensions of wildlife.

Anna Lydia Svalastog is Professor of Psychosocial Work at Østfold University College, Norway, and associate professor at the Centre for Research Ethics and Bioethics at Uppsala University, Sweden. She holds a PhD in history of religion from Uppsala University, and is a permanent faculty member (docent) at the Faculty of Humanities, Umeå University, Sweden. Her research focuses on past and present conditions and frames for individual and societal life, health and knowledge. She is Chair of the Navigating Knowledge Landscapes Network, focusing on health and knowledge in the digital society, and she is a member of the expert committee for the H2020 PANELFIT, a project working on the European general data protection regulation (GDPR). She has co-edited books on myth and rites in the field of religious studies, Sámi studies, psychosocial work, and health and the digital society.

Sondre Svalastog was an associate professor in tourism at Mid-Norway University (previously Lillehammer University College). He holds a PhD in geography of economics from NHH in Bergen, Norway, and a BA and MA from the Norwegian University of Agriculture, now NMBU, Ås, Norway. From 1972 until his death in 2012, he worked at the Department of Tourism at Lillehammer College. From the early 1970s, Svalastog collaborated with Swedish researchers in the field of tourism, as Sweden at that time already had an established competence in the field. He collaborated with most of the contributors to this book for decades, including McNeil, whom he met when McNeil was a parish priest in Lillehammer. Svalastog had a group of collaborative partners that he involved as guest lecturers. The group included Dahle, Arne Næss, Nils Faarlund, Sigmund Kvaløy Setreng and Hjalmar Hegge. During Svalastog's sabbatical in 1984–85 in Edinburgh, Anthony

Travis was the director of the department. They became close colleagues and friends. The collaboration with Travis introduced Svalastog to Marko Koščak with whom he conducted several excursions in Norway, as well as in Slovenia. From 1986 to 1989, Svalastog had a three-year scholarship at the Norwegian School of Economics where Ole Arnljot Strømme Svendsen quickly included him and his family in the academic and cultural life of Bergen. The two became close friends and colleagues for the rest of Svalastog's life. Svendsen was also the key discussant during Svalastog's last work, his report on the decline of the tourist industry in Norway. From the 1990s, Svalastog was a regular guest speaker in Sweden. In addition to being called to Kalmar University, and occasional work for Uppsala University, he was, several times, invited to Karlstad University by Lars Aronsson. From 2006 to 2007, Svalastog was an invited researcher at Umeå University with Dieter Müller as his host. Svalastog had national and international colleagues and friends in the areas of cultural heritage and museums, ethnology, and cultural studies. The importance to him was expressed through his membership and involvement in the International Council for Monuments and Sites (ICOMOS). Throughout his academic life, he worked closely with Statistics Norway and with colleagues he had met during his time at the University of Agriculture, as a student or, later, at the Department of Rural Sociology.

Anthony S. Travis was a professor who started his career as a town and country planner, after completing his wartime schooling in New Zealand, postwar schooling in Wales and, later, at the University of Manchester in England. After his town planning career, culminating in his first planning chair in Edinburgh and, later, chairs in urban studies research, in Birmingham, and tourism, in Glasgow, he became a renowned worldwide as an international tourism expert, consulting for the European Union, the UN World Tourism Organization and the Pacific Area Travel Association.

Foreword

The Art of Science, or for the Love of Life

Anna Lydia Svalastog and Dieter K. Müller

This book presents Sondre Svalastog's academic contribution to the tourism industry together with contributing case studies. Svalastog's work is well known in Norway as well as to contributors to this book, to his fellow members of the International Council for Monuments and Sites (ICOMOS) and to his many friends and colleagues around the world. Though parts of his main work have previously been published in English, Svalastog left behind an outline of a book based around this study and formed a research group that met for the first time in 2005. The aim of the group was to develop case studies exemplifying the usefulness of Svalastog's conceptual and theoretical work. Following his death, his daughter, Anna Lydia Svalastog, contacted the group and together they decided to complete the project.

Key agents throughout the process have been Anna Lydia Svalastog, Dieter Müller and Brian McNeil. As editors, Anna Lydia Svalastog and Dieter Müller would like to acknowledge and thank Inland Norway University of Applied Sciences (formerly Lillehammer University College) for allowing them to use Svalastog's text. They would also like to thank Brian McNeil for the work he has done on translating and correcting all text not written by Sondre Svalastog.

Thanks also go to the Norwegian Non-Fiction Writers' and Translators' Association (NFFO) for their generous economic support, enabling the completion of this book.

Further thanks must be given to Sondre Svalastog's widow and lifelong partner, Borgny Svalastog, for engaging Tim Challman to translate his Norwegian texts and Tarjei Svalastog for the necessary improvement of Sondre Svalastog's models, tables and maps, and for organizing the photographs. Thanks also to all the case study authors for their contributions, enthusiasm and support, reflecting their appreciation of Sondre Svalastog and this project.

Sondre Svalastog came from the southern tip of the Hardanger mountain plateau of Norway. He loved his family and his garden and he cherished friendship. He was deeply connected to traditional handicrafts and ways of living and used these to influence his home and the places where he lived. He was raised in the countryside with strong traditions of handicrafts, timber houses and self-sustaining households. He kept the past close to the present and regarded his ancestors as merely a heartbeat away. He understood previous generations – who they were, where they came from and how their lives evolved. He regarded the past as inseparable from the present and believed everything around us to be interrelated with and inseparable from nature. The beauty of taste and function contained in generations of local knowledge and sustainability were especially dear to him.

Sondre was greatly loved and is deeply missed. He will always be vividly remembered as the man who enjoyed a tasty meal, drinking red wine, giving long, well-prepared speeches at feasts and dinner events, and who danced like Zorba in a celebration of life.

16 February 2020

Preface[1]

Sondre Svalastog

This work had a lengthy genesis. When I drew up my methodological and theoretical report on 'Attitudes of the Local People to Tourism: Reasons – Models – Hypotheses' in 1979 (15 pages + 87 pages of appendices), I little guessed that this would be the starting point that has guided most of my research work down to the study that I present here, which is a kind of summing up of a project with three partial phases.

The first phase (1979–1984) was primarily a period of fieldwork, when empirical evidence was sought in order to test concepts, models and hypotheses that were integrated into the methodological and theoretical report. Thematically, work in this period was organized under the title 'Local people's attitudes to tourism'.

The second phase (mid-1984–mid-1989) concentrated on resource-theoretical questions. This was the natural continuation of the first stage, because my work in charting local people's attitudes to the tourist industry revealed that very many of these attitudes, whether conscious or unconscious, were based on the competition for free goods, the competition between various sectors to obtain resources that have a value on the market and are profitable in resource-based tourism.

In the third phase, the primary concern was questions about localization in the macro context. This is why I have discussed the general theory of production, the theory of the international division of labour and the theory that explains which factors are central in order to become internationally competitive. I believe that this will be a very central challenge for planning in the future: How can one make a well-founded link between the general level of prices and costs in a region (or a nation) and the types of products and structures of production that are capable of achieving sufficient competitiveness on the international market?

The bibliography indicates those to whom I am intellectually indebted; but besides these authors, there are several individuals who have meant a great deal to me in the course of this work, by commenting on specific parts of the project, by drawing my attention to important scholarly literature, or by showing great interest in the work and encouraging me to continue with it.

In the initial phase, I am especially grateful to the anthropologists Arne Martin Klausen (University of Oslo) and Ulla Wagner (Ethnographic Museum, Stockholm); to Arne Martin Klausen, first and foremost, because he informed me about several recently published studies by his colleague Ulla Wagner and encouraged me to contact her about my project. This proved to be highly useful.

[1]This preface was part of Sondre Svalastog's PhD thesis published in Norwegian. The English translation is by Brian McNeil.

Ulla Wagner, who at that time held a post at the University of Stockholm, followed the first phase of the project in its entirety through a series of work sessions and seminars. I am also deeply grateful to the economist Fritz C. Holte (Norwegian University of Life Sciences), the political scientist Ottar Brox (University of Tromsø) and the ecophilosopher Hjalmar Hegge (University of Oslo) for comments and discussions of individual 'papers' in this phase.

From the first phase down to the second phase of the project, I should like to mention especially the geographer Brian S. Duffield, who had done a great deal of work in the 1970s and 1980s on charting and evaluating resources for tourism and recreation. In the early 1980s, he was the director of the Tourism and Recreation Research Unit at the University of Edinburgh. Brian S. Duffield invited me to Edinburgh, where I was a visiting researcher in 1984/85. A little earlier, the geographer Anthony S. Travis, at the University of Birmingham, had published a study entitled 'Physical impacts: trends affecting tourism: managing the environmental and cultural impacts of tourism and leisure development' (*Tourism Management*, December 1982). This article was the key that enabled me to acquire a practicable concept for tourism resources. I was able to visit Anthony S. Travis in Birmingham while I was in the UK and we have developed our academic contacts since then.

During the course of this work, I also had contact with the sociologist Børge Dahle (Norwegian School of Sport Sciences), the sociologist Jens Kristian Steen Jacobsen (my colleague here at University College), the geographer Lars Aronsson (University of Karlstad) and the geographer Martin Price (Environmental Change Unit, University of Oxford). These contacts have meant a great deal to me.

At a later stage in the second phase, the project was formalized as a doctoral dissertation at the Norwegian School of Economics in Bergen. In December 1987, a dissertation committee was appointed. Its members were the geographer Peter Sjøholt (Norwegian School of Economics), the economist Arnljot Strømme-Svendsen (Norwegian School of Economics) and the sociologist Helge Solli (Norwegian University of Life Sciences). This formalization of the work was very important for me. The dissertation committee has always encouraged me and has always been willing to discuss individual problems and the overarching organization of the material.

In financial terms, by far the most important support was the three-year doctoral grant I received from the Norwegian School of Economics (August 1986–July 1989). I have also received subsidies to cover my expenses at various stages of the project from the following bodies (I refer to them by the names they held at the time): the tourism section of the Ministry of Transport; the Ministry of the Environment; the District Development Fund; the National Team for Tourism in Norway; the Council for Research in Social Planning; and the three local districts that were involved – Øystre Slidre in Oppland, and Bykle and Risør in Aust-Agder.

The graphic designer Kjell Helge Sjøstrøm and the computer consultant Kjell Totland (both at the Institute for Geography at the Norwegian School of Economics) helped me, with kindness and great patience, to prepare the diagrams and the chart material for the final report. I wish to thank Nina Karlsen, Torild Tveit and Linda Østby at the library in the Norwegian School of Economics for the immense amount of work they have undertaken in examining all the references to scholarly literature in the dissertation and in the nine partial reports that preceded it. The infrastructure that the Norwegian School of Economics made available to me was decisively important, and I am very grateful that I was able to draw on this assistance until the work was finished. In the final phase, the consultant Trude Israelsen helped me greatly in writing the text and eliminating errors.

Sondre Svalastog
Lillehammer, March 1993

Image of Grungebru in the early 20th century. In addition to the mountain hotel, Nils Svalastog continued to run a large annual horse-trade market that could bring together up to 300 horses. The horse traders paid a fee per horse. *Private photo: Anna Lydia Svalastog*

Grungebru tourist hotel and farm. Picture from the 1960s. Sondre's father, Nils Svalastog, bought Grungebru in the late 1930s. Together with his wife Gunhild, they ran a tourist hotel (*turistheim*). The hotel was combined with farming and a small sawmill. In the early/mid-1970s, they built a new barn and house on the other side of the nearby river and sold the tourist hotel to the Youth Hostels Association. Two years later their daughter bought it back, together with her husband Oscar Johan Garnes, and continued to run it as a youth hostel. *Postcard from Grungebru; Anna Lydia Svalastog's collection*

From the very early days of the Grungebru hotel. Nils Svalastog bought the land in the late 19th century and built and opened the mountain hotel in the early 20th century. Nils Svalastog was an uncle of Nils Svalastog (b. 1912), Sondre Svalastog's father. *Private photo: Anna Lydia Svalastog*

Grungebru, after it had been expanded and reshaped as a tourist hostel. *Postcard from Grungebru; Anna Lydia Svalastog's collection*

Remembering Sondre Svalastog

Brian McNeil

I first met Sondre Svalastog in the late summer of 1989 when he moved with his wife Borgny to Lillehammer. Borgny was already a Catholic and they became regular worshippers in the Catholic Mariakirken, where I was parish priest from 1988 to 1990. The congregation at that time was extremely small – only about 90 persons scattered over several counties of central Norway. The family made a tremendously important contribution to the life of the parish.

We maintained contact after I left Norway to become a parish priest in Munich. I received him into the communion of the Catholic Church in Lillehammer on 28 September 1996. He attended the Silver Jubilee of my ordination in Munich on 3 May 2005 and I celebrated his requiem mass in Lillehammer and buried him in Grunge Cemetery on 9 March 2012.

I learnt a great deal from him as a person and as a scholar. Let me begin with the second point. More than 20 years ago, Sondre asked me to translate his doctoral dissertation into English, but this project (for reasons I no longer remember) never got off the ground. Over the years, however, I translated a large number of texts for him, some for publication and others for circulation among his colleagues. He suffered from the drawback that all Norwegian scholars experience: although the Norwegian languages Bokmål and Nynorsk are rich and nuanced, and perfectly capable of expressing every facet of human experience and knowledge, there are simply too few persons outside of Scandinavia who can read and appreciate a text in Norwegian. This means that somebody has to translate their work into English (and if the Norwegian scholar has written a text in English, this must be revised, since it is notoriously difficult to write idiomatically in a foreign language).

Sondre's writings are marked by a great conceptual precision and clarity in their formulation. They reveal him to be both an actor on the international scholarly stage and one who is anchored in a very specific local context, to which he pays great attention. This combination of the international and the regional, where each illuminates and enriches the other, presupposes not only a breadth of knowledge but also the ability to see patterns and interconnections. Translating his texts opened my eyes to understand much that I had only superficially observed.

Sondre was a man of great culture. In the beautiful memorial book *Requiem* (edited by Per Bjarne Bøym and Ivar Roger Hansen, Røros, 2015), his grandson, Tarjei Svalastog, recalls breakfasting in the garden of the house in Lillehammer:

> Lofty conversations were conducted over the table, and at regular intervals, either Grandfather raced off to his study, or Grandmother down to her studio, to fetch several kilos of books that might enlighten the discussions. After an hour at the breakfast table, the coffee cups were perched on the edge of the table in the shadow of heaps of books that would rise 30–40 cm into the air.

Tarjei also speaks of Sondre's great dedication to work and of his bravery after his cancer was diagnosed. Obituary notices sometimes speak of a person who has died 'after a great fight'. Sondre, too, put up a great fight; his spirit and his scholarly eros were undimmed.

His religious convictions were unostentatiously but very deeply held all his life. After he became a member of an international church, he continued to display his interest in its regional and local dimensions. I remember a number of discussions with him about the future of Bjerkebæk, the house in Lillehammer where the Catholic novelist Sigrid Undset spent many years of her life.

One cannot speak of Sondre without speaking in the same breath of Borgny, the great love of his life. In my sermon at the requiem, I quoted Isolde's words in the second act of *Tristan and Isolde*, when she sings 'Is not our love called Tristan *and* Isolde? This sweet little word "*and*"!' I suggested that we experienced something similar when we found it so natural to speak of 'Borgny *and* Sondre'. I also suggested that the grief we felt at Sondre's death was, in a sense, the price we pay for love. We hear every day of deaths but we mourn only when we have lost a person who meant much to us.

If he meant so much to us, this is surely because we meant so much to him. He evoked affection and love because that was what he gave to others. There is a beautiful expression in Norwegian that sums it up: *Vi lyser fred over hans minne* ('We call down peace upon his memory').

Childhood in Vinje

Youth

Studies at Ås

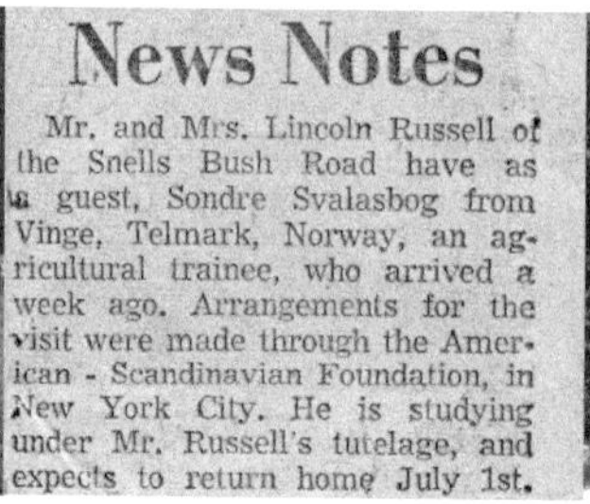

News Notes

Mr. and Mrs. Lincoln Russell of the Snells Bush Road have as a guest, Sondre Svalasbog from Vinge, Telmark, Norway, an agricultural trainee, who arrived a week ago. Arrangements for the visit were made through the American - Scandinavian Foundation, in New York City. He is studying under Mr. Russell's tutelage, and expects to return home July 1st.

Studies in New York State and England

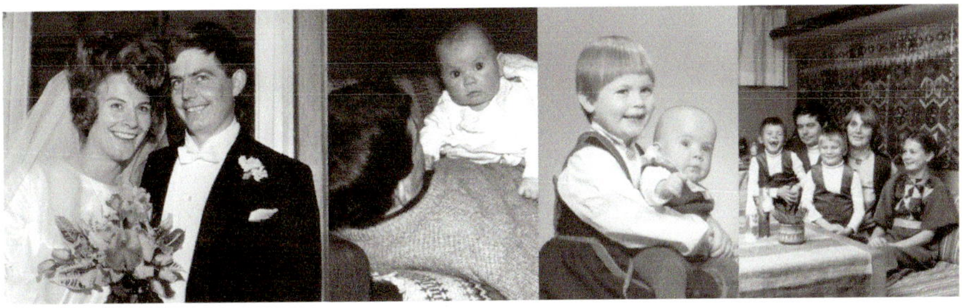

Marriage

Scholar

Family

Family

Photographic Credits

p.ii – Sondre in duffel coat and home-knitted winter hat in the mid-1970s. *Private photo: Anna Lydia Svalastog*
p. xx – all photos, except one, owned by Anna Lydia Svalastog

First row (Childhood in Vinje), left to right:

Sondre at Grungebru, about one-and-a-half years old, 1941; Farm work with siblings (*left to right:* Tone Svalastog Garnes (b. 1946), Astrid Svalastog (b. 1953), Tore Olav Svalastog (b. 1950), *in front:* Sondre Svalastog (b. 1940); Sondre testing a new machine for cutting grass (photo: Sondre Hustveit)

Second row (Youth), left to right:

Speartime skiing (Sondre on right), mid-1950s; Fishing in the river at Grungebru (Sondre on left, Tore Olav in front); Sondre, 1958 or 1959, from a collection of photos of the class at Tomb School of Agriculture

Third row (Studies at Ås), left to right:

Rune Gerhardsen, Prime Minister of Norway, giving a speech for the students at the Norwegian College of Agriculture (early to mid-1960s); Surveying (early to mid-1960s); Sondre with fellow students at the Norwegian College of Agriculture (Sondre second from left)

Fourth row (Studies in New York State and England), left to right:

From Sondre's period of agricultural practice in the USA, 1962 (Sondre on left); Newspaper notice about his agricultural practice in the USA, 1962; From Sondre's scholarship period in the UK during his studies at the Norwegian College of Agriculture, 1963

Fifth row (Marriage), left to right:

Borgny and Sondre, wedding day, 16 April 1966; Sondre with daughter Anna Lydia, 1967; Anna Lydia and brother Nils (b. 1970); Family photo featuring youngest child, Olav Kyrre (b. 1973)

Sixth row (Scholar), *left to right:* Sondre defends his PhD, 1993; Sondre in his home office; Sondre on excursion, 2005.

Seventh row (Family), left to right:

Sondre with parents and three younger siblings. *Left to right:* sister Gunla (b. 1942), mother Gunhild (b. 1915), Sondre, father Nils (b. 1912), brother Tore Olav (b. 1950), *front:* sister Tone (b. 1946). Sondre with his parents after he defended his PhD, 1993; Sondre with wife Borgny, 1997; Sondre with grandson Bendik (b. 1995), 2011.

Eighth row (Family), left to right:

Sondre with grandson Tarjei (b. 1990), 2011; Sondre's children: *from left to right:* Nils (b. 1970), Anna Lydia (b. 1966), Olav Kyrre (b. 1973); Sondre at his timber cottage in Vinje, Telemark, 2011

1 Introduction: A Resource Approach to Tourism

Anna Lydia Svalastog, Ian Jenkins and Dieter K. Müller

Values and Sustainability

This book represents the work of Sondre Svalastog and his ambition to establish a conceptual and theoretical framework for research and planning within the tourism industry. His research can be characterized as transdisciplinary: a methodological and theoretical fusion of natural science, social science (in particular sociology), economics (in particular economic geography), cultural history and philosophy. Svalastog was formally trained in the field of agricultural sciences, sociology and economics.

He studied at the Norwegian University of Agriculture, which included a scholarship to visit England in 1964. In 1966, after finishing his studies, he worked at the Department of Rural and Regional Sociology at the Norwegian University of Agriculture. Thus, from the very beginning, his focus on tourism was established with his approach being both interdisciplinary and international.

Svalastog was an academic who represented key concerns of the twentieth century. Both in his thinking and his work he always considered the wider context of impacts, ethics and sustainability on society, culture and nature and viewed them as an integrated and working whole. At the beginning of the 21st century, these perspections, insights, conceptual tools and theoretical frameworks have been shown to be more important than ever before, both in the field of the tourist industry and beyond.

Throughout his working career, Svalastog was involved in societal planning, primarily at county level. His fieldwork focused at the level of municipalities, since the complexity of tourism and its potential and challenge vary from city to countryside, as well as from one rural municipality to another. Svalastog strove to keep together all the different aspects of societal life, including resources generated in the past and their likely future consequences, industry-related potential as well as social impacts, economic values and the prosperity, health and well-being of all stakeholders, not least the people and areas who were affected as intended or unintended hosts for tourism and leisure activities.

A key concern for Svalastog was the importance of local conditions. These, he argued, needed to be fully understood and included in any analysis, in order to generate tourism that is both productive and sustainable. For example, it might be assumed that small, local villages around the mountain plateau of Hardangervidda constitute one single region, but this is not the case. Several municipalities still use seasonal mountain holding areas for their animals, which limits how the tourist industry is able to use such landscapes; additionally, infrastructure including roads and public transport, varies. In addition to understanding the local conditions and

 Tourism as a Resource-based Industry: Based on the work of Sondre Svalastog (eds A.L. Svalastog, D.K. Müller and I. Jenkins)
DOI: 10.1079/9781800621480.0001

resources, there needs to be knowledge about the type of tourist attracted to particular regions so as to maximize potential and production, ensuring that both the community and people benefit. Successful planning in the tourist industry requires knowledge about social, cultural and natural changes.

The importance of climate change is a key area of knowledge in the tourist industry. From the early 1980s, Svalastog included climate change as necessary and important for the sustainable planning of tourism. This part of his research met with resistance and attempts were made to restrict the dissemination of his research to the public. The controversy revealed how short-term economic interests, both in academia and the tourist industry, can be strong in their determination to silence uncomfortable research. This particular controversy concerned climate change, but the general point is that in a constantly changing world where the tourist industry is large and economically important, research on tourism needs to be in a process of constant renewal of risk analysis, oriented towards society, culture and nature at the same time. To ensure sound planning within the tourist industry, Svalastog insisted on the need for research-based knowledge. It becomes evident that Svalastog's work was highly innovative, especially his recognition of the role of climate change on a destination's resources. It was clearly ahead of its time, and even in the early 1990s anthropogenic climate change was only beginning to be recognized as having future impacts on tourism. The impact of climate change encompassed one of the key elements of Svalastog's work, that of knowledge-based resources. These were only recently recognized as an important element in the success of a destination. Success will not happen if a destination does not have the knowledge and skills necessary for adapting to climate change.

In the early 1970s, Svalastog was asked to help establish academic studies of the tourism industry in Norway, which was a relatively new field. He was trained in science and sociology and continued to collaborate and work with researchers in the field of cultural studies (anthropology, ethnology and folklore). He also undertook advanced studies in economics, and in his PhD in economic geography, the abstract nature of academic disciplines became obvious. As tourism was a new academic field, it lacked the academic roots needed to ensure complex and well-elaborated concepts and theoretical understandings. His answer was to actively invite key academics as guest teachers and to participate in his research groups. Most importantly, he ensured that philosophy and the history of ideas were included.

Svalastog collaborated with leading Norwegian researchers engaged in the philosophical critique of positivism. In Norway, this post-war debate emphasized how academic thinking and research are always part of society, reflecting the values and understanding of the society in which they are produced. Consequently, in Svalastog's work, immaterial culture is just as important as material culture. In addition to his academic work and work in societal planning, he operated within the frames of UNESCO and was a voting member and the principal investigator of the Norwegian research group ICOMOS (International Council for Monuments and Sites). His understanding was that critical reflection, together with conceptual and theoretical clarity, is a necessary starting point for all inquiries, as well as for societal planning.

In present-day global and (post-)industrial society, tourism is a key industry; however, it also presents ethical challenges and risks that mirror social categories and economic inequalities, stemming from an industrial and premodern period. Of particular concern is the (re-) production of class and gender, as well as colonialism. Svalastog contributed to the ICOMOS tourism charter, which was an explicit answer to these challenges.

Because of its complexity – the interrelated societal levels (global and local), resources (natural, cultural, social, economic) and plurality of values (economic values; ethical values; international norms, values and laws) – tourism as a field of academic investigation requires an integrated transdisciplinary approach. This book presents an approach to the tourist industry based on knowledge of natural resources, social organization and its institutions, together with cultural traditions and know-how. Furthermore, the book also acknowledges the responsibilities of research to support humanity and nature and its importance for societal planning and development.

Inquiry-driven Transdisciplinary Research

Svalastog walked a long academic path, enabling him to properly understand different academic landscapes including their research history and methods. This eagerness to extend his areas of competence was motivated by two factors: his research on tourism and sustainability and also the needs that were generated from this field of investigation (including his ethical stance on tourism). He understood humans and nature, past and present, material culture as well as the intangible. His perspective on life was seen through an understanding of it being relational, dynamic and constantly changing. The gaze that he created was of humans as agents, responsible for the consequences of their actions; as a society we should actively strive to investigate and acquire new and relevant knowledge as well as revisit old assumptions.

Knowledge presupposes precise definitions and well-elaborated theoretical perspectives. Over the years, it became a key challenge for Svalastog to develop a toolbox of interrelated concepts. These included key analytical tools for identifying and analysing challenges and possibilities related to societal planning and implementing sustainability within the tourism industry. This book presents the results of this particular part of Svalastog's work. The case studies are meant to demonstrate the usefulness of his toolbox and to develop it further with reference to particular contexts of the industry.

Svalastog emphasized the importance of complex and context-related knowledge, based on well-established research methodologies. His stance implied a double perspective: academic knowledge and history represented reliable knowledge, which was built on a reflective model; at the same time he emphasized the importance of local knowledge and tradition (non-academic knowledge) as crucial for culture and society and thus also for tourism. He valued highly, and gave his life to, academic knowledge production, in the field of systematic investigation-driven transdisciplinary work. At the same time, he repeatedly emphasized the tentative character of academic borders and the constant need to push, transcend and alter these to gain new insights.

Sustainability in Present Society

Research-based knowledge is a prerequisite in societal planning, a fundamental principle for governmental governance and a key for legitimizing policy actions. International agreements and international collaboration refer explicitly to research and science as key ingredients in development and sustainability. The importance of Svalastog's work lies in his insistence on the importance of integrated knowledge and the relationship between academic knowledge, local knowledge, governance, planning and ethics.

Since the end of World War II, ethics and human rights have been in the forefront of governance, regulation and planning. Ethics and human rights have been transported into all areas of international regulation, institutionalized through the UN and its different bodies such as UNESCO and ICOMOS. These organizations of the UN have, in turn, been in control of charters directed to specific social areas; for example, the ICOMOS charter on tourism. Guidelines have been developed, ratified and implemented to ensure a fair and natural, social and culturally sustainable global society. However, times seem to be changing once again. During the 2015 Paris meeting on climate change, the representatives of global societies reached a point where conflicts between cultural and natural sustainability risked pulling people apart. So far, the international guidelines have been included in the planning, but now key actors, including Norway, suggest stripping away crucial arguments regarding culture and human rights in order to achieve the two-degree goal. This change in political direction is supposedly validated by references to natural and social science data. From Svalastog's perspective, this is a contradiction in terms and therefore undermines long-term sustainability. In his thinking, research and cultural knowledge are fundamental principles for planning, although in a fused transdisciplinary way. Svalastog understood the wider context of sustainability incorporating the four pillars: Environmental, Economic, Societal and Cultural, as well as the multi-disciplinary nature of developing tourism. Much public understanding of sustainability, even today, appears solely environmental but Svalastog's work included both the social and cultural aspects of destinations and resources.

Research is a key validator for the new priorities in present-day global society, as highlighted in the Paris agreement of COP21 and, hopefully, more fully acknowledged in COP26. The game-changer is not research and knowledge itself but the way politics and economics work together in ways that create conflicts of interests and fragmentation. Because of this, in key areas of societal disputes, research seems to have been misused to trump arguments regarding ethics and human rights. Seen from Svalastog's perspective, the conflicts are created out of a lack of insight into life's fundamental interrelatedness and how interrelatedness, not conflict, constitutes the real challenge: How do we protect 'the spirit of place'? How do we protect life and ensure sustainable planning and development?

In Norway, the so-called positivism-critique has been most important. In Svalastog's work, research represents norms and values and his stance must be understood as a critical challenge to the positivist approach. In present society, reference to research is made in ways that tend to under-communicate or even hide norms and values, and reference to research-based planning seems to re-establish the former naïve positivist stance on research. In Europe, this new wave of positivism has been around for a few years, not least due to new technology-driven research.

To conclude this part of Svalastog's perspective, there are two research-related challenges embedded in our understanding of the relationship between research, planning and a sustainable society: (i) although a study may be scientific, due to its scientific methodology, it can be less than helpful if it excludes perspectives or material related to the question(s) and location(s) that are being investigated; and (ii) science is tentative and the scientific method does not of itself produce results that answer questions of governance. What makes a particular study relevant depends on a variety of contextually defined factors and, in general, it demands more thorough studies than those that are planned for by planners. Furthermore, questions about ethics and human rights need to be integrated and constantly revisited in the planning process. In the end, governance is politics based on priorities and preferences and should not be excused by simplified references to research, as if we do not have a say in how we live our lives and organize our societies. Since this book is meant to give a presentation of the main components needed to undertake an analysis that can be used for the sustainable planning of tourism, the intention is not to be an end-point, but to create an academic reflection relevant to planning and politics and to fuse further conceptual and theoretical works on tourism and planning.

Localizing Tourism Geographies

Tourism has been a field of scientific investigation and research since the beginning of the 20th century (Hall and Page, 2014). This applies to the Nordic countries too, where early examples can be found from the 1930s onwards. Nevertheless, tourism as an area of research and education remained marginal until the 1970s when newly established university colleges in Norway and Sweden focused on this still relatively neglected field. Institutions of higher education in tourism were established in order to supply an educated workforce for the regional tourism industry, while research often remained a somewhat marginal activity.

At that time, tourism was not regarded as a major discipline at Nordic universities, and even elsewhere tourism departments were rare. Hence the challenge for those young scholars hired at the new institutions was intense, since curricula and study materials had to be newly developed – little could be taken from elsewhere. Moreover, scepticism towards studying tourism and leisure was even more widespread then than it is today and government resources for research were usually not provided, at least not to the same extent as at the established universities (Butler, 2004).

Against this background, it was equally important that tourism developed not only as a field of education but also of research. In Nordic countries, as elsewhere, geography was an important discipline, producing many of the scholars who were active at the new tourism departments (Müller, 2010). This development has continued, and today a majority of articles published in *Tourism Geographies*, the journal most clearly committed to tourism geography research, come from departments that are not identifiable as geography departments (Müller, 2014).

This book is inspired by the work of one of those pioneers, Sondre Svalastog. After degrees in rural economy (1966, Norwegian Agricultural University) and planning (1971, Norwegian Agricultural University), Svalastog was appointed, in 1972, Lecturer in Local and Regional Planning at the Oppland University College, which had been established a year earlier (since 1994, University College Lillehammer; and since 2017, Inland Norway University of Applied Sciences). Even then, he was committed to discussing the conditions for tourism development in the Norwegian regions. He continued and deepened these efforts despite the scarce resources available for research at that time, finally earning a PhD in economic geography from the Norwegian School of Economics in Bergen in 1994. His thesis, 'The localization of tourism: on the analysis of resources, spatial distributions and local adaptation' (*Lokalisering av reiseliv: om ressursanalyser, den romlige fordeling og lokal tilpassing*), summarized a comprehensive research effort covering 19 case-study areas and involving interviews with more than 2400 people.

A Resource Approach to Tourism

Tourism is usually understood as a system linking tourists' origins and destinations, while acknowledging the role of transportation to convey tourists to locations. Moreover, the tourism system is located within an environment that also establishes preconditions for the development of tourism (Hall, 2005). Although all parts of this system may be affected by the development of tourism, it seems that, in particular, the destination has garnered most attention among researchers. This also applies to Sondre Svalastog, who engaged in a core study of tourism geographies, i.e. how and why destinations develop. As an economic geographer, Svalastog was inspired by and departed from Zimmermann's Theory of Resources (1951). Accordingly, resources are not fixed entities; rather, they come into existence and are made by people who attach meaning to them. In Svalastog's (1998) understanding, this meant that destinations, too, are in the making. The necessary production process comprises various aspects. Already Clawson *et al.* (1960) offered a classification of resources for recreation (Hall and Page, 2014):

- The undeveloped recreation resources: landscape and its natural features like lakes, rivers, mountains etc.
- Private recreation resources: second homes and other privately owned resources
- Commercialized private recreation resources: shopping malls, theme parks, stadiums, resorts etc.
- Publically owned recreation resources: sports facilities, national parks etc.
- Cultural resources: public and private, libraries and arts etc.
- Professional resources: administrative resources comprising financial support systems and organizational structures, and management including planning and research.

However, the resources considered often included land, workforce and capital. Clawson *et al.* (1960) also claimed that available recreational resources are usually sufficient for satisfying a local recreation demand. However, in cases where resources are thought of as tourist attractions, outstanding resources are needed in order to achieve competitiveness. A good example of this is the growth in adventure tourism which, since the 1970s, has increased significantly in popularity; then it was simply seen as outdoor recreation. Although not specifically mentioned by Svalastog, he obviously recognized and understood the potential for what is now known as adventure tourism. He identified knowledge as a key component for success, which is supported by other authors also recognizing this within the Norwegian adventure tourism sector (Løseth, 2017).

Furthermore, Svalastog's research, published in 2010, highlighted the importance of the adventure tourism market to Norway. He identified the need for developing this product, which, according to other research, represented at least 18% of Norwegian tourism. Svalastog notes:

> The production of activity-based recreation[2] is much more important for the tourism sector than would be suggested by its share in wealth creation, if that is considered on its own. It is in the production of activity-based recreation that the main emphasis of the non-material increase in value must lie. (Svalastog, 2010, p. 6)

Once more we see that Svalastog is able to clearly identify a market which has now become a global player in attracting tourists and which Norway has capitalized upon.

As if to underscore the development of Norwegian adventure tourism, many adventure companies market the country as an adventure destination. It has even featured in the British press, as the UK is one of Norway's main markets (Stone, 2017). The *Global Report on Adventure Tourism* notes: 'The trend is far-reaching. In 2011, 79% of tourism boards reported that the adventure tourism private sector had begun to emerge and/or grow in their destination' (UNWTO, 2014, p. 22), and it further adds: 'Around the world, destinations gear their taglines and messaging to appeal to adventure travelers: Norway capitalizes on its towering fjords and glaciers with the slogan "Powered by Nature"'(UNWTO, 2014, p. 22). This, in many ways, confirms Svalastog's research and understanding of knowledge-based tourism development, showing that, currently, adventure tourism is a significant product in developing a successful global tourism destination.

Svalastog drew a distinction between place-specific resources, comprising Clawson *et al.*'s undeveloped recreation resources, and manipulated man-made resources. In this context and inspired by Travis's (1982) work on the management of tourism impacts and Duffield's (1984) attempt to bridge the gap between tourism studies and economic geography, Svalastog pointed to the role of immaterial aspects and the fact that resources are culturally determined. Thus, he showed in his thesis that assessments of tourism impacts were contingent upon the degree of dependency on tourism development (Svalastog, 1994). The latter notion implies that two other dimensions become important: awareness and knowledge.

Svalastog recognized that, in particular, tourism development in Norway was suffering from the development of the Norwegian oil economy, resulting in high costs for the production of tourism. Hence, despite the expected development of mountain tourism in Norway in the 1960s, it turned out that international tourism in Norway had actually decreased (Svalastog, 1988). The expansion of an undeveloped physical recreation resource turned out to be insufficient to attract tourists from the international market. Svalastog blamed the development on price levels in Norway for causing this failure. Because of the high price level, the tourist product was uncompetitive in an international market despite the indisputable attractiveness of Norwegian scenery. In order to deal with the situation, he claimed that a greater input of knowledge in the production of tourism in Norway was the way forward (Svalastog, 1992, 2010). This would enable a more professional approach and help to produce an internationally viable touristic product. A review of current Norwegian tourism identifies that the recreation market has indeed expanded with a considerable increase in the availability of adventure tourism products.

This insight entailed another aspect of Svalastog's thinking on geographies: a conclusion drawn from the assertion of lack of competitiveness of Norway's tourist product was that geography and geographical knowledge should be responsible and applied in order to benefit communities and societies. From this knowledge platform, Svalastog became involved in ICOMOS. ICOMOS is a non-governmental organization associated with UNESCO, engaged with the protection of monuments and heritage sites. Moreover, ICOMOS functions as an advisory commission for the implementation of the World Heritage Convention of UNESCO. Svalastog was active in the Norwegian chapter of ICOMOS and for many years chaired its committee on culture and tourism. He also represented the committee in international meetings and was active in the promotion of ICOMOS and its ideas.

Tourism in economic geographies

As current authors sometimes forget (Ioannides, 1995, 2006; Hall, 2013; Müller, 2014), a gap between tourism geographies and mainstream geographies was already noted in the 1980s (Duffield, 1984). Svalastog's work, too, can be seen in this context. Trained as an economic geographer, Svalastog was obviously concerned with the drifting apart of tourism and geography. Consequently, his thesis (1994) has to be seen not only as an attempt to measure local perceptions of tourism but also as an application of economic (resource) theory to tourism and a test of whether this approach was applicable in supporting aspects of its production. His focus on the role of knowledge was relatively unusual at the time but was to become very fashionable in research on economic geography and regional development only a few years later (see, e.g., Maskell and Malmberg, 1999; Cooke, 2001; Bathelt *et al.*, 2004).

Until today, this focus on knowledge and skills within tourism geographies is limited (Baum, 2007; Xiao and Smith, 2007) and Cooper (2006) argues that the tourism industry has been poor in adopting knowledge, at least partly because of gaps and distance between academia and industry. Moreover, the focus is often on skills and knowledge management at a micro level and within the individual firm (e.g. Moscardo, 1997; Aitken and Hall, 2000). Often, skills development is seen as a means of economic restructuring (Thomas and Long, 2001), particularly within tourism in developing economies (e.g. Kaplan, 2004; Liu and Wall, 2006; Baum and Thompson, 2007). However, little effort has been made to assess the role of knowledge for competitiveness within the tourism industry and the connection between knowledge and the provision of tourism resources. Research on national parks may be the exception; here a greater awareness of the connection between the experiential dimensions of tourism, tourism management and conservation is discussed relatively intensively (Eagles, 2002). In general, Svalastog's work is seldom replicated and addressed, providing a further reason to make his work, often originally published in Norwegian, accessible to an English-speaking audience.

Outline of the book

This book provides the reader with a glimpse of Sondre Svalastog's work and it is structured in two parts. Following the introduction, selected parts from Svalastog's PhD thesis are presented in Part 1. Some of the texts are reprints, while others are available in English for the first time. The second part features a number of contributions from former colleagues and friends of Svalastog. These are original contributions and, in fact, some are provided by pioneers of tourism geographies, such as Anthony Travis. However, the texts aim at providing an empirical context for Svalastog's theoretical elaborations. Some of these texts are almost fragments providing insights into the original thinking of innovators in the discipline, while others conform to modern academic texts.

References

Aitken, C. and Hall, C.M. (2000) Migrant and foreign skills and their relevance to the tourism industry. *Tourism Geographies* 2(1), 66–86.

Bathelt, H., Malmberg, A. and Maskell, P. (2004) Clusters and knowledge: local buzz, global pipelines and the process of knowledge creation. *Progress in Human Geography* 28(1), 31–56.

Baum, T. (2007) Human resources in tourism: still waiting for change. *Tourism Management* 28(6), 1383–1399.

Baum, T. and Thompson, K. (2007) Skills and labour markets in transition: a tourism skills inventory of Kyrgyzstan, Mongolia and Uzbekistan. *Asia Pacific Journal of Human Resources* 45(2), 235–255.

Butler, R. (2004) Geographical research on tourism, recreation and leisure: origins, eras and directions. *Tourism Geographies* 6(2), 143–162.

Clawson, M., Held, R. and Stoddart, C. (1960) *Land for the Future*. Johns Hopkins University Press, Baltimore, Maryland.

Cooke, P. (2001) Regional innovation systems, clusters and the knowledge economy. *Industrial and Corporate Change* 10(4), 945–974.

Cooper, C. (2006) Knowledge management and tourism. *Annals of Tourism Research* 33(1), 47–64.

Duffield, B.S. (1984) The study of tourism in Britain: a geographical perspective. *GeoJournal* 9(1), 27–35.

Eagles, P.F. (2002) Trends in park tourism: economics, finance and management. *Journal of Sustainable Tourism* 10(2), 132–153.

Hall, C.M. (2005) *Tourism: Rethinking the Social Science of Mobility*. Pearson, Harlow, UK.

Hall, C.M. (2013) Framing tourism geography: notes from the underground. *Annals of Tourism Research* 43, 601–623.

Hall, C.M. and Page, S. (2014) *The Geography of Tourism and Recreation: Environment, Place and Space*, 4th edition. Routledge, London.

Ioannides, D. (1995) Strengthening the ties between tourism and economic geography: a theoretical agenda. *The Professional Geographer* 47(1), 49–60.

Ioannides, D. (2006) Commentary: the economic geography of the tourist industry: ten years of progress in research and an agenda for the future. *Tourism Geographies* 8(1), 76–86.

Kaplan, L. (2004) Skills development in tourism: South Africa's tourism-led development strategy. *GeoJournal* 60(3), 217–227.

Liu, A. and Wall, G. (2006) Planning tourism employment: a developing country perspective. *Tourism Management* 27(1), 159–170.

Løseth, K. (2017) Knowledge development in adventure tourism businesses – the influence of serious leisure. *Annals of Leisure Research* 21, 575–591.

Maskell, P. and Malmberg, A. (1999) Localised learning and industrial competitiveness. *Cambridge Journal of Economics* 23(2), 167–185.

Moscardo, G. (1997) Making mindful managers: evaluating methods for teaching problem solving skills for tourism management. *Journal of Tourism Studies* 8(1), 16–24.

Müller, D.K. (2010) Turismgeografi eller turism? Reflektioner om utvecklingen i ett ämne. *Geografiska Notiser* 68(2), 61–67.

Müller D.K. (2014) 'Tourism geographies are moving out': a comment on the current state of institutional geographies of tourism geographies. *Geographia Polonica* 87, 299–312.

Stone, A. (2017) 10 reasons to visit the happiest country on Earth. *The Telegraph.* Available at: www.telegraph.co.uk/travel/destinations/europe/norway/articles/norway-best-adventures-and-activities/ (accessed 3 October 2021).

Svalastog, S. (1988) Tourism in Norway's rural mountain districts twenty-five years after the Mountain Planning Team's report. *Norsk Geografisk Tidsskrift* 42(2/3), 103–120.

Svalastog, S. (1992) Tourism in a changing Norway. *Norsk Geografisk Tidsskrift* 46(2), 109–116.

Svalastog, S. (1994) *Lokalisering av reiseliv: om ressursanalyser, den romlige fordeling og lokal tilpassing*. Høgskolen i Lillehammer, Lillehammer, Norway.

Svalastog, S. (1998) The nature of resources for tourism and recreation. Working paper 68/1998. Høgskolen i Lillehammer, Lillehammer, Norway.

Svalastog, S. (2010) *Om den raskt avtakende konkurransekraft i norsk reiseliv: En produksjonsteoretisk analyse*. Høgskolen i Lillehammer, Lillehammer, Norway.

Thomas, R. and Long, J. (2001) Tourism and economic regeneration: the role of skills development. *International Journal of Tourism Research* 3(3), 229–240.

Travis, A.S. (1982) Managing the environmental and cultural impacts of tourism and leisure development. *Tourism Management* 3(4), 256–262.

UNWTO (2014) AM Reports, Volume 9 – *Global Report on Adventure Tourism*. UNWTO, Madrid.

Xiao, H. and Smith, S.L. (2007) The use of tourism knowledge: research propositions. *Annals of Tourism Research* 34(2), 310–331.

Zimmermann, E.W. (1951) *World Resources and Industries: A Functional Appraisal of Availability of Agricultural and Industrial Resources*. Harper, New York.

2 Tourism in Local Communities

Sondre Svalastog

Introduction

In the first two chapters, I discuss the choice of topic for this dissertation, along with the hypotheses and models that existed when I began the project 'Integration of Tourism in Local Communities'. Furthermore, I will utilize the results of the interview survey 'Local Populations' Attitudes towards Tourism' to test key elements in my theoretical foundation for the concept of the tourist's encounter with the local community. Altogether, this constitutes a natural first stage. Various works, some of a general social science nature and some specifically related to tourism,[1] have given me reason to speculate that there may be:

- Important distinctions between *different local communities*, primarily urban and rural. Based on general studies of a social science nature, I was led to conclude that integration of tourism in urban areas (as a challenge to public planning) was relatively simple. This is partly because tourism is predominantly a result of modern society, or, more specifically, tourism as a mass phenomenon is an urban phenomenon. Tourism in cities is about modern people meeting other modern people. Also, the relative numbers of permanent residents compared to visiting tourists in an urban context are in a totally different proportion when compared with sparsely populated rural areas. If my introductory statements appear to be tentative, it is due to the fact that nearly all integrative studies (studies of an anthropological or sociological nature) have been linked with rural areas or smaller holiday towns. Larger cities with large business-related traffic would not have been the object of comparable attention.
- Important distinctions can be made between *different kinds of tourism/travel*, mainly between business-related travel on the one hand and various forms of leisure-related tourism on the other. The expectation that this distinction exists is due principally to the fact that business travel is carried out within the home place's social ties and control and in interaction with people who most frequently share much of the same cultural and experiential background. Additionally, production to meet the business-related demand is sheltered. The producers therefore do not experience the same price pressure on their products. Production is less seasonal than ordinary leisure-related demand and therefore generates employment that is essentially different from that of companies producing for the leisure-related market.

The term *local community* will be used here to refer to social systems that can be delimited on

DOI: 10.1079/9781800621480.0002

a graduated scale up against other social systems. The term *local community* will often denote a municipality, an area of a large city or a rural community. Larger units in the hierarchy will include counties and greater society (the nation), and at the opposite end of the spectrum, households. The term *local community* means, in practice, the lowest entities in the context of *public holistic planning.* In other words, this means entities in which the residing citizens (on the electoral roll) are entitled to communal services, such as schooling, and services provided by the health and social care agencies for which public planning has the objective of providing for a good quality of life. Under the term *local community*, I initially want to emphasize formal affiliation and the consequences that various measures might have on the manner in which common services will be, or can be, provided to individual members of the community. I will place less emphasis on the actual social network. In principle, I choose to follow the tradition that stresses the key difference between social networks in traditional rural communities, as opposed to the social networks that we find in urban areas. The key difference consists in the geographical range. Several of the close social relationships, in the tradition that I choose to follow, can be anticipated to be found within local, rural communities, whereas most city dwellers have their close social relationships spread over a larger geographical area.

My notion of the objective of *public holistic planning* is the same as that of the planning theoretician Faludi, who held that 'On a certain level, planning in the public sector may always be regarded as "social" in that it is geared towards the achievement of social ends – whatever that may mean! Practically, even the planning of private corporations has a strong "social" element. They, too, require internal adjustment of ends similar to the socio-political process and they, too, have to adapt to their social environment' (Faludi, 1978, p. viii).

As a starting point, I chose to build up to a study of encounters between various local communities and different types of tourism. Individual studies of encounters between visiting travellers and permanent residents throughout the 1970s have been quite frequent studies among anthropologists and sociologists. When I began my work in 1979, many 'case studies' were already readily available. The most comprehensive systematization of individual studies in the startup phase was the work entitled *Host and Guests: The Anthropology of Tourism*, by Smith, published in 1977. This work was twofold. Firstly, it contained a large number of 'case studies' from various parts of the world. The case studies were categorized according to principal cultural areas. Specifically, they were categorized as follows:

- Nascent Tourism in Non-Western Societies
- Tourism in European Resorts
- Tourism in Complex Societies

In addition, this work had an introductory paragraph in which the editor attempted to generalize on the basis of the 'case studies' contained in the book. I will return to the results of this section and will include an elaboration in further sections including: 'On the grouping of travellers according to characteristics and traits', 'Tourist typologies' and 'Choice of typology'.

My own study will also revolve around encounters with travellers and tourists, but because it is so directly linked to public planning, I will have to work with all types of travellers. This means, firstly, that the *business traveller* must be included on an equal footing with the *leisure tourist.* To obtain a grasp of the entire field of tourism in a more explicit manner than has commonly been the case, I shall also divide the leisure-related part of tourism into a *commercial* element (the normal tourist travel for pleasure) and a *non-commercial* element (visit-related tourism). Naturally, sociologists and anthropologists were primarily interested in the commercial part of leisure tourism because it was *more exotic* and more clearly represented the encounter between groups having *different cultural backgrounds* than between families, relatives and friends visiting on holiday. To the extent that economists were interested in the field and developed 'case studies', these were predominantly macro studies at national level, or they were economic impact studies at the regional level. The introductory chapter will be primarily used to develop a typology that can be gainfully suitable for my study.

Moreover, I anticipated that it must be possible to ascertain important causal relationships between the business traveller and the leisure tourist, which could be used to develop a more universally valid theory for 'Integration of Tourism in Local Communities'.

The Business Sphere and the Leisure Sphere

In developed countries, work and leisure have remained, for most people, two separate spheres. The work sphere has become a thoroughly investigated and regulated collective matter, whereas the leisure sphere, similar to religion, belongs to the private domain. Sociologist Ramsøy describes these two spheres as follows: '... the leisure situation is the time and activities that the individual resorts to on his/her own, makes his/her own choices, in spontaneous activity having no other purpose than immediate or short-term personal satisfaction. Leisure time is the individual's holiday from society' (translated from Ramsøy, 1978, pp. 659–660).

Significant portions of holiday and leisure-related traffic arise precisely out of people's desire to abandon their daily social commitments for a limited period. A striking example of this is the investigation of social anthropologist Wagner that was published under the telling title *Out of Time and Place: Mass Tourism and Charter Trips* (Wagner, 1977). In this study, Wagner analyses the home situation and the holiday situation of Swedish charter tourists travelling to The Gambia.

The part of this work that is most applicable for our purposes is the documentation of how this category of tourist abandons their social commitments at home and assumes an 'out of time and place' situation during their holiday stay. After the holiday is over, they return to the social relations that they had temporarily abandoned for a period of time.

Business-related tourism is different, however. There is reason to believe that it is unproblematic in a social sense, because it consists of travellers who move within their daily social system with routine procedures and work that is to be done among like-minded individuals. The business traveller, in other words, avoids the social control of the community. Furthermore, we know that the business traveller has the capacity to pay and is a modest consumer of *goods without a market price*. The manner in which tourism belongs partly to the leisure sphere and partly to the business sphere is illustrated in Fig. 2.1.

If one groups these in terms of ascending significant economic categories, the leisure traveller is a *consumer*, while the business-related traveller is a *supplier*. The business traveller stays

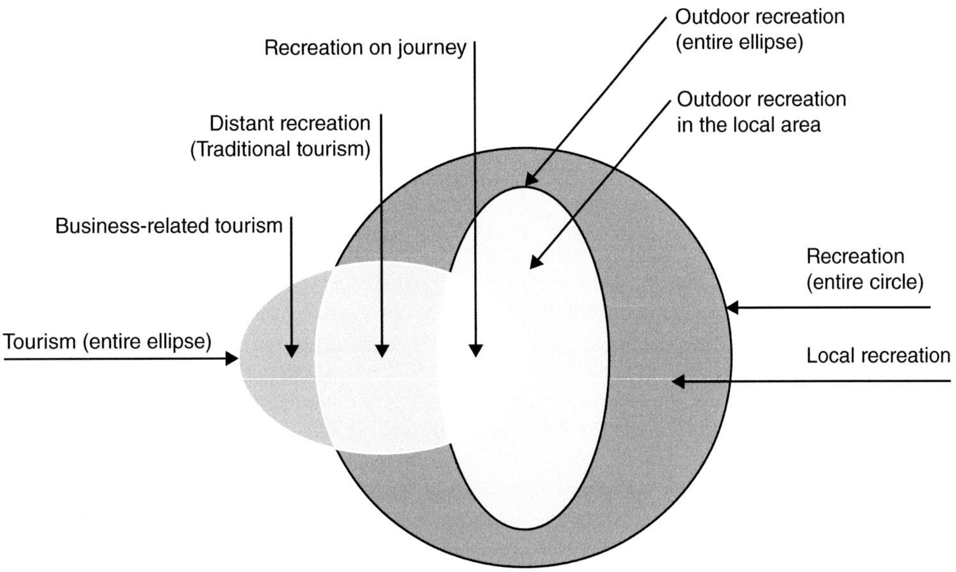

Fig. 2.1. Business-related tourism and leisure-related tourism. (Adapted from Flognfeldt, 1976, p. 16[2])

at the place visited as a supplier on behalf of the place visited, while the leisure traveller stays there by virtue of his/her financial capacity to be a consumer. This consumption can target *goods/resources that have a market price*, but it often also targets goods/resources without a market price – what we might call *free goods*.

On the Grouping of Travellers According to Characteristics and Traits

Tourism as a phenomenon has been analysed, delimited and split up into distinctive subgroups or different traits and objectives. Sociologist Erik Cohen, in his work 'Who is a tourist?: a conceptual clarification', establishes a very interesting distinction, sociologically speaking, between the generic term *traveller* and what he argues can be termed a *tourist* (Cohen, 1974, pp. 527–553). The tourist in this analysis is one who travels voluntarily during his/her free time and one who is a stranger to the place visited. This entails a difference for both the business traveller and the leisure traveller who visits relatives and friends, has a holiday home or returns to a holiday area frequently; these fall outside the notion of a 'tourist' – they are merely travellers. I wish to borrow from this work primarily the premise that the leisure traveller who is a stranger to the place visited is, sociologically speaking, something other than the leisure traveller who returns to familiar nature and culture.

The historian Boorstin (1964) uses modern mass tourism to postulate that modern people mainly seek 'pseudo-events'. This may also be an interesting dimension to make use of, but it is not directly applicable to the typological outcome I wish to attain. The dimension appears interesting in this respect, chiefly in connection with a theory of resources. In such a conjunction, therefore, it may be profitable to return to this perspective.

Contrary to Boorstin, sociologist MacCannell views the tourist and the tourist's journey as our era's variant of a traveller on a pilgrimage or a quest for the sources of various aspects of our modern society (MacCannell, 1973, p. 593). This is also an interesting perspective on the tourist and the tourist's journey and it is the antithesis of the essence of Boorstin's notion of the modern tourist. The perspective is immediately interesting in connection with a theory of resources, as was Boorstin's perspective. My hypothesis on this point is that it is neither one nor the other; we have tourist groups that are attracted by 'pseudo-events' and other tourist categories that consciously seek authenticity in their holiday experiences.

In a later work, the anthropologist Turner focuses on the holiday stay as temporary freedom (holiday) from the social setting of one's home place along with the beneficial interaction that this represents for the tourist (Wagner's work, to which I referred earlier, places the main emphasis on the cultural consequences that such social freedom has in relation to the recipient society). In this respect, Turner has used the term *communitas* to stress that the holiday situation may offer totally different opportunities for spontaneity, 'personal wholeness' and social closeness (Turner, 1974, pp. 80–154). This may be an interesting dimension to bear in mind, because indirectly it may tell us something essential about the tourist or about the adaptation of certain tourist groups in the receiving area.

The works to which I have briefly referred above each grasp, in their own way, interesting dimensions between types of travellers, but for each of these works the point of departure has differed from the starting point I have for mine. What I wish to basically achieve is segmentation, or tentative division, of the traveller phenomenon based on the rationale for why they left their home place and what they are seeking at the place they visit.

At the inception of this project, three works seemed to support this premise. These are:

- Vilhelm Aubert (1969): Chapter 6 'Tourism', from *The Hidden Society*, which was published in Norwegian for the first time in 1969 (most recent edition 1985). The first edition was published already in 1965 under the same title *The Hidden Society* but did not include this tourism chapter;
- Erik Cohen (1972): 'Towards a sociology of international tourism' from *Social Research* 39, 164–182; and
- Valene L. Smith (1977): 'Introduction' (pp. 1–14) in *Hosts and Guests: The Anthropology of Tourism* (new edition published 1989 – no notable changes to the Introduction).

Aubert's work comes the closest to what I seek to achieve, illustrated by his introductory statement:

> A classification for sociological analysis should be rooted in traits that pertain to the stress and pressure that exists in society's social structure and that the tourists flee from, as well as in the social properties that characterize the niches offered to the tourist in host countries and in the local communities where he is received.
> (Aubert, 1969, p. 117)

Here Aubert places the sociological focus on the permanent dwelling place and the place visited. Furthermore, Aubert considers everyone to be on a leisure-oriented trip away from their permanent dwelling place. What interests Aubert, then, is everyone, not just the sociological issue of 'Who is a tourist', as we find in the above-mentioned work by Cohen (1974). The distinction that Cohen establishes here becomes extremely important if one wants to study the encounter between various local societies and different kinds of travellers. On the other hand, Aubert does not include those travelling for business-related purposes. To apply this to Fig. 2.1, Aubert principally analyses all travellers within the recreation circle, but business-related travellers fall outside. The other two authors do not place so much emphasis on the society that the tourists leave; they concentrate primarily on the tourist at the destination visited and their role as visitor.

Tourist Typologies

In this section, I look more closely at the three typologies that directly focus on the encounter between the traveller and the local community. To begin with, I will review Aubert's tourist typologies, continuing with a review of the typologies developed by Cohen and Smith.

Aubert posits six main types of tourists (leisure travellers), placing great emphasis on highlighting the tourist's role in, and relationship with, the society he travels from as well as the tourist's role and adaptation to the society he visits for a limited period of time. The academic strength of Aubert's work lies in his systematic examination of what individual tourists leave behind and how they adapt socially and materially to the place they visit. Hereafter, the keyword *seek* will be used as a commonality describing what the tourist seeks to leave behind and what he seeks to find.

a. The outdoor tourist

This is the tourist who turns his holiday into a total change of situation. First, the outdoor tourist flees daily social obligations and everyday conformity. Second, he seeks physical and mental challenges. The holiday trip requires meticulous planning and the participants are thoroughly resourceful people in terms of financial, social and intellectual status. The participants thus come from the political, academic and business elite in highly industrialized parts of the world.

b. The anti-tourist

This is the leisure traveller who does not wish to be perceived as a tourist, but rather wants to be a social and cultural consumer in the local society where he temporarily stays. The anti-tourist category principally comprises two groups, each of which encounters the local society with entirely different expectations. The greater proportion in this group are people who have moved away from their home place and who seek to return to their roots during holidays.

Socially, culturally and interactively, they easily fit into and become a part of the community during their stay. This segment of the group therefore falls outside the sociological definition of a tourist such as it is delimited in Cohen's 1974 work. The other group comprises visitors who want to adapt socially and materially to the local community during their stay. Because the stay is of limited duration, this adaptation may be only partial, but in our context the desire to adapt is what is essential. The visitor does not knowingly demonstrate differing values, morality and attitudes; the visitor strives for cultural respect.

c. The emigrant tourist

This is the tourist who 'spends a significant portion of his time away from his native country because he is against certain aspects of its societal structure' (Aubert, 1969, p. 118). The motive here is more to get away from aspects that one intensely dislikes; for example, a too-strictly regulated society, a society that is not free, a too-transparent society etc. Socially speaking, these are often resourceful 'non-paying passengers' who are not part of an obligating social teamwork either at home or in niche destinations abroad. 'Since their most salient motive is to get away, they seek only a comfortable and cheap place for their tourism, without worrying much about the receiving society's qualities aside from the prices, the taxes and the weather' (Aubert, 1969, p. 118).

d. The socializing/sightseeing tourist
'For this category of tourist, the journey *per se* is not so important. Spatial movement could well have been replaced by another type of instrument to bring him away from everyday social relations and over to others' (Aubert, 1969, p. 136). What is essential for this voluminous, large group is the social relationships with fellow tourists, or the quasi-social opportunities that pictures and experiences offer at home following the trip. Nature and culture are secondary but, nevertheless, important backdrops.

e. The expert/connoisseur
This is a category of tourists who have special interests that they nurture during their holiday. 'They are interested in an aspect of the receiving society and its population, such as folklore, music, handicrafts and applied art, or simply their everyday customs and the social and economic aspects of the host country' (Aubert, 1969, p. 134).

The expert often possesses academic or professional qualifications. He spends his holiday attending to interests that have been neglected due to insufficient time as a result of work pressures. The holiday becomes an 'extension' to accommodate important personal interests in the tourist's everyday life. (For this term, see Parker, 1965, pp. 73–74).

f. The sun worshipper
This is the tourist that primarily chooses an area because of climate (dry and warm for summer activities and stable, dry and sunny winter climate for sports and recreation), 'combined with a desire to live the good life' (Aubert, 1969, p. 117). The sun worshipper seeks fellowship with other tourists. The charter tourist travelling to southern holiday areas is a classic example. In Norway, this means tourism to the major winter sports resorts, and during the summer to areas along the southern coast of Norway. For most people, their everyday life role is far removed from their role as tourists.

Cohen's somewhat later work is restricted purely to leisure travellers who cross national borders (international tourism) and, of these, what he calls 'modern tourism'. Historically and globally, inclusive products such as pilgrimages, visits and leisure-related trips within one's own country fall outside this. Nevertheless, the work provides important concepts and building blocks for my own work due to the terms that are developed and the designated contexts.

In Cohen's work, the following two hypotheses are key: firstly, he identifies pertinent points relating to the degree of domestic and foreign tourism and considers these 'on the basis of their relationship to both the tourist business establishment and the host country (Cohen, 1974, p. 164). By making a main point out of the way the journey is organized, he highlights the types of travel that provide opportunities for a genuine encounter with *local nature and local culture* and through this, opportunities to learn; and, on the other hand, the types of travel that provide very limited possibilities for real encounters and learning. Cohen formulates the hypothesis as follows:

> There is a continuum of possible combinations of novelty and familiarity. This continuum is, to my mind, the basic underlying variable for the sociological analysis of the phenomenon of modem tourism. The division of the continuum into a number of typical combinations of novelty and familiarity leads to a typology of tourist experiences and roles.
>
> (Cohen, 1974, p. 167)

On this continuum, Cohen establishes a major distinction between what he calls 'the institutionalized form of tourism' and 'the non-institutionalized form of tourism'. Cohen visualizes what he wants to highlight by saying that the institutionalized form of tourism (what he calls mass tourism) transpires in a bubble of domestic environment, but non-institutionalized tourism lacks such a bulwark between it and *local nature* and *local culture*. This is a concept I will apply in the continuation of the work on this project.

The other hypothesis in Cohen's work is the following:

> The sad irony of modem institutionalized tourism is that, instead of destroying myths between countries, it perpetrates them. The tourist comes home with the illusion that he has 'been' there and can speak with some authority about the country he has visited. I would hypothesize that the larger [the] flow of mass tourists becomes, the more institutionalized and standardized tourism becomes and consequently the stronger the barriers between the tourist and the life of the host country become.
>
> (Cohen, 1974, p. 174)

We will not directly use this hypothesis in the present project, but it offers an important perspective on the encounter between the tourist and the local community and therefore is interesting to bear in mind. The typology that Cohen presents includes the following four tourist types:

Institutionalized tourism

The collective mass tourist

The entire journey is programmed from home down to the smallest detail for the tourist. All stops (experiences), meals, overnights, guiding etc. are scheduled beforehand. All events along the way are interpreted via an intermediary such as a guide, the captain, the driver, a tourist brochure etc. 'In this respect, the quality of the mass tourist's experiences approaches that of vicarious participation in other people's lives, similar to the reading of fiction or the viewing of motion pictures' (Cohen, 1974, pp. 169–170).

The collective mass tourist mainly corresponds with the following groups in Aubert's typology:

- the sun worshipper (the most prevalent);
- the sightseeing tourist; and
- the expert/connoisseur (to a certain extent, primarily the city-oriented trips).

The individual mass tourist

This type of tourist also outsources most of his planning since 'all of his major arrangements are still made through a tourist agency' (Cohen, 1974, 167–168). The daily legs of the journey are largely set from home and often include vouchers purchased in advance for overnight accommodation or reservations made from home. The sights that are included are the grand and the spectacular. They are frequently of a historical nature and mirror aspects of another era. 'His excursions do not bring him much further afield than do those of the organized mass tourist' (Cohen, 1974, p. 168).

The individual mass tourist will primarily correspond with the same tourist types from Aubert's typology of the organized mass tourist, that is:

- the sightseeing tourist (the most prevalent);
- the sun worshipper; and
- the expert/connoisseur (to a certain extent).

Non-institutionalized tourism

The discoverer

The discoverer is a financially, intellectually and physically resourceful traveller. The trip itinerary is planned to avoid the established tourist places, and he plans the content of the trip on his own. However, he is not willing to give up the personal comfort he is used to having at home. His contact with local nature and culture is direct, but it is intellectually judgemental and classificational. 'He remains somewhat detached, either viewing his surroundings from an aesthetic perspective or seeking to understand the people on an intellectual level' (Cohen, 1974, p. 175). The discoverer is the precursor to the mass tourist.

The discoverer corresponds, to a large degree, with what Aubert classifies as an expert/connoisseur.

The drifter

This is the tourist who seeks a stay on the premise of accepting the local society's material and cultural conditions. 'He shuns any kind of connections with the tourist establishment and considers the ordinary tourist experience phony' (Cohen, 1974, p. 168). The drifter allows the trip to unfold as he goes along; he visits the unfamiliar motivated by the desire to experience, and he tries at times to become a part of the local subculture. He shares some traits in common with the wandering journeymen that we know from the Middle Ages, since neither the itinerary nor the destination are known beforehand. When necessary, he takes random odd jobs along the way. On the other hand, the drifter has a curious trait in that he is recruited from the upper social levels in society and he returns to that stratum when he is ready to take social responsibility and have a family.

Essentially, the drifter corresponds with the anti-tourist in Aubert's typology.

The last segmentation work we will include is the 1977 work of the American anthropologist Smith. Unlike Cohen, she does not explicitly define what is included in her notion of the term 'tourist', but she exemplifies it throughout the article, drawing on domestic tourism (for example, Las Vegas, North Carolina, the French Alps, First Nation areas in the south-western regions of the USA, Alaska and Hawaii where 'domestic tourism' is highly tangible) as well as international tourism. As in Cohen's work, she limits her analysis to 'modern tourism'.

The typology developed by Smith encompasses five categories and seven tourist types. The categories are types of 'umbrellas' incorporating common features in experiences. The five categories are: ethnic tourism, cultural tourism, historical tourism, environment-related tourism and recreational tourism. The categories primarily reveal what the holiday targets, in terms of content, while the division into tourist types 'can be drawn, accounting for their numbers, their goals, and their adaptations to local norms' (Smith, 1977, p. 8). Principally, this is an interesting point of departure for our academic purpose and in a pedagogical sense the author succeeds in developing a model that visualizes in an illustrative manner the key significance in the encounter between various types of tourism and the local society. We will transform and utilize the figurative representation as we continue. The problem in our context is that several of the tourist types in Smith's work encompass very different tourists. The types of tourists are inconsistent in terms of motivation, needs, drivers and social attitudes. For example, Smith allows mass tourism (Smith, 1977, p. 11, Fig. 2) to include:

- hitchhikers
- inter-railers
- individual travellers
- participants in company trips

and they purchase all kinds of accommodation, so that 'they fill up hotels of every category, pensions, and hostels' (Smith, 1977, p. 10).

We can admit that this large group shares this trait in common with the situation at home, but, as pointed out earlier, we assume that the holiday trip is often an individual's holiday away from the social control of the home place; the holiday traveller assumes roles that comprise other values, norms and social attitudes than those of everyday life. Furthermore, we believe that the same person, over time, goes in and out of various tourist roles. This suggests that Smith's typology is not well suited to my purposes. Aubert's typology is far more consistent in terms of values, norms and social attitudes. I believe, for example, that the western middle-class generates quite different types of tourists, such as:

- the sightseeing tourist
- the expert/connoisseur
- the anti-tourist.

Smith's basic hypothesis is that 'a touristic typology can be drawn, accounting for their numbers, their goods, and their adaptions to local norms' (Smith, 1977, p. 8). The typology is structured as follows:

The discoverer

According to Smith, this tourist is rare. They share traits more in common with anthropologists undertaking fieldwork than with traditional tourists. They adapt to local material and cultural conditions while at the same time utilizing many western aids. The discoverer is resourceful; this type of holiday is costly and demands resources.

The elite tourist

The first two types of tourists closely resemble one another. The primary distinction between them is that elite tourists 'differ from explorers because they are "touring"' (Smith, 1977, p. 9).

They are few in number and they are financially, intellectually and physically resourceful people. This type of tourist is recruited from the upper classes in industrialized parts of the world. They often hire local guides, bearers and other locally recruited helpers.

The deviant tourist

Smith includes two subgroups under this tourist type. One subgroup comprises those who try to dissociate themselves from the 'tourist crowds' while the others 'seek to heighten the excitement of their vacation by doing something beyond the norm' (Smith, 1977, p. 9). They travel frequently and go long distances; costs and energy consumption are high despite them being very capable of adapting to local situations.

The adventurer

This is the tourist that considers himself, as do those in his circle of acquaintances, to be an 'adventurer'. However, objectively speaking, he is not. The discoverer, the elite tourist and the deviant tourist are all types who may share adventurous qualities (if by that one means curiosity about the unexplored and unexpected). The tourist journey is largely set up by the travel agency at home with the tourist carrying out the trip more as an observer.

The advent of mass tourism

Mass tourism is characterized by a steady stream of tourists travelling to universally well-known tourist destinations. They expect a high degree of western adaptation. The group is dominated by participants from the middle classes who are snobbish in the sense that they demonstrably refuse to adapt materially or in terms of values. A large number of these tourists are individual travellers.

Mass tourism

According to Smith, this type of tourism 'is built upon middle-class income and values' and 'they expect a trained, multi-lingual hotel and tourist staff to be alert and solicitous to their wants as well as to their needs' (Smith, 1977, p. 10). As an attempt at constructing a generic term for this group, Smith coined the phrase 'The "tourist bubble" of Western amenities'. This finds its parallel in the phrase, 'his familiar "environmental bubble" ... that we find in one of the works referred to above' (Cohen, 1972, 166–167).

Charter tourists

Large numbers and standardization are the most salient traits. Here, inclusion of domestic elements from home are most prevalent and everything is transformed via the guide, travel brochures, the personnel that the tourist agency has stationed at the hotel etc. The encounter with local culture and local nature is from a distance.

Choice of Typology

As mentioned above, I shall avail myself of Smith's idea in the figurative representation of the encounter between different forms of tourism and local society, but I will incorporate Aubert's typology in terms of leisure-related tourism. I will then supplement Aubert's typology with my own group, encompassing business travellers.

Before we include tables illustrating the segmentation works I have reviewed, I wish once again to emphasize that all tourist categories are equally important to include, since the main challenge of this project is integration of 'tourism in local communities'. The following global relationship is therefore important to keep in mind:

- 'The volume of domestic tourism worldwide is estimated by the WTO to be about ten times greater than international tourism'[3] (Pearce, 1988, p. 95).
- 'A more detailed study by the WTO of variations in purpose of visit by major regions in 1979 showed vacation travel accounted for almost three quarters of all international movements' (Pearce, 1988, p. 37). The calculated shares are 71% holiday, 13% business trips, 16% other, of which the classic pilgrimages are one of the largest items in this category.
- In 1979, based on the calculations above, international modern leisure tourism represented approximately 7% of the total global travel volume.

In Table 2.1, I illustrate the differences between the four segmentation theories referred to above, in relation to what they identify.

What we now have as a point of departure is, in principle, a typology that encompasses both leisure-related and business-related tourism. Furthermore, it is a universal typology since it comprises both national and international tourism. It can be used for other time periods than our own (it is not limited to modern tourism as with Cohen and Smith) and it can be used in cultural contexts other than our own (because, among other things, religious journeys are included). If we now sum up the distribution into various tourist types, illustrated with the aid of the 'triangular model' found in Smith's work, we obtain Fig. 2.2.

Fig. 2.2 primarily visualizes the following:

- The three unproblematic forms of leisure-related tourism in an integration context: the outdoor tourist, the anti-tourist and the emigrant tourist. All correspond essentially with Cohen's main group, 'the non-institutionalized form of tourism'.
- I have placed business-related tourism in a zone that indicates the main form is essentially unproblematic. Strong control mechanisms are the main reason for this. In a global context, however, we see that it can represent different suppressive elements, for example through material luxury consumption and as a market for prostitution. Therefore, this main form is placed in a transitional area.
- The remaining forms of leisure-related tourism, namely the sightseeing tourist, the expert and the sun worshipper, are forms that correspond with Cohen's main group, 'the institutionalized forms of tourism' (mass tourism). These forms of tourism are problematic in terms of integration. Several important factors make this true, partly because of the volume of tourism itself, partly because of the lack of ability to understand local nature and local culture and partly because local nature and local culture are of interest only as backdrops to the social life that the traveller wishes to practise. Local nature and local culture therefore become instru-

Table 2.1. What the various segmentations identify.

MAIN FORMS		OGILVIE* 1933	AUBERT 1969	COHEN 1972	SMITH 1977
Profession-dependent tourism	Internationally	●	○	○	○
(Market-based tourism)	Domestic	●	○	○	○
Pleasure and experience tourism	Internationally	●	●	●	●
(Resource-based tourism)	Domestic	●	●	○	●
Visiting tourists	Internationally	●	●	○	○
(Network-based tourism)	Domestic	●	●	○	○
● With	○ Omitted				

* Ogilvie's is not a segmentation work like the others are, but more a discussion of 'who is a tourist' in an economic sense (Introduction p. 5–7). What we need to be particularly aware of here is that all four authors take 'the tourist' as a point of departure, and that we find a main distinction between Ogilvie; on the one hand, discussing the phenomenon in an economic sense, and the other three using and perhaps discussing the term in a sociological or socio-anthropological context. Ogilvie; includes tourism linked with the leisure sphere as well as tourism linked with the business sphere. The other interesting aspect to note is that Ogilvie's conceptual content largely corresponds with the content that is recommended by The World Tourism Organization (Baron, 1989, 11–25).

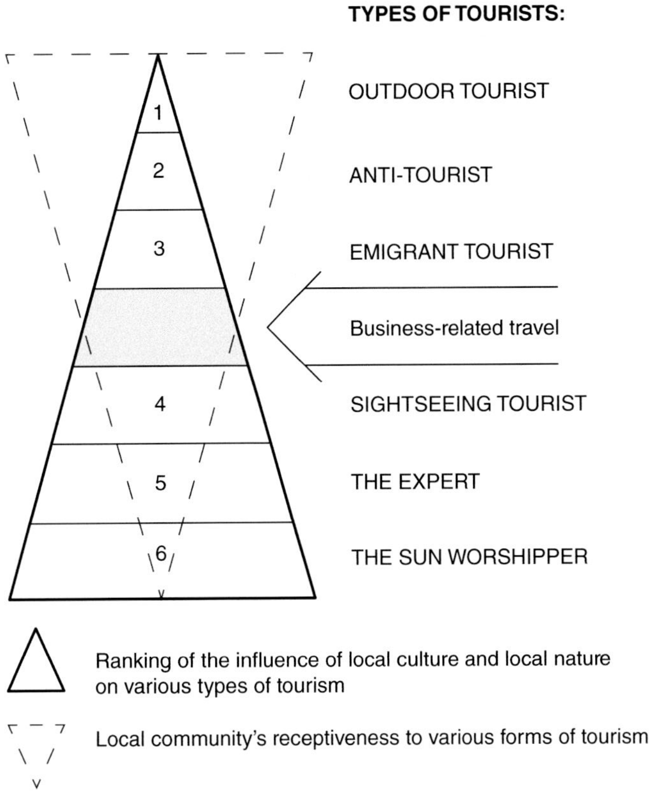

Fig. 2.2. Different tourists and differences in local integration.

ments utilized at the discretion of the traveller and not necessarily with respect or as local society would view them.

The Encounter between Local Society and Visitors

It is possible only up to a point for consequences and attitudes to be two sides of the same coin. This is because myths and erroneous interpretations, or quite simply wrong information, can be important elements in the attitudes that exist in specific geographical regions or population groups, with regard either to one concrete question or to an entire area. Besides this, aggressive exogenous or endogenous memories can alter the consequences, thereby opening the way to changes in attitudes. If an industry changes, we may expect that the consequences for society will also change.

Any matters pertaining to social interactions are beset by a lack of objectivity and, in some circumstances, rationality. This is because each individual and the groups they comprise have a number of factors that affect perception:

- value perception
- aligned interests
- perception of reality (familiarity with existing alternative choices and ongoing societal changes).

The emphasis on these is a direct consequence of my understanding of the purpose of public planning and that:

- up until now, in the context of planning (in the public as well as the private sphere) we have concentrated on material conditions; and
- we have justifiable grounds to assume that the most decisive attitudes are of a social nature and/or are (culturally) based on values.

In my analysis, I include both material and non-material consequences. To ensure a full understanding of the dimensions involved, I establish a methodological approach to my analysis that looks at five interrelated areas (Fig. 2.3).

Why are these five major areas of consequences given equal rank *a priori*? The answer is because each major area is so essential that it can individually supply sufficiently strong arguments for changes through planning. Or else, on the contrary, it can support such strong objections that it prevents a change of priorities through planning.

We have now reached the stage of establishing the development of a final model that can be universally used to describe the encounter between various local societies on the one hand and various forms of tourism on the other (Fig. 2.4).

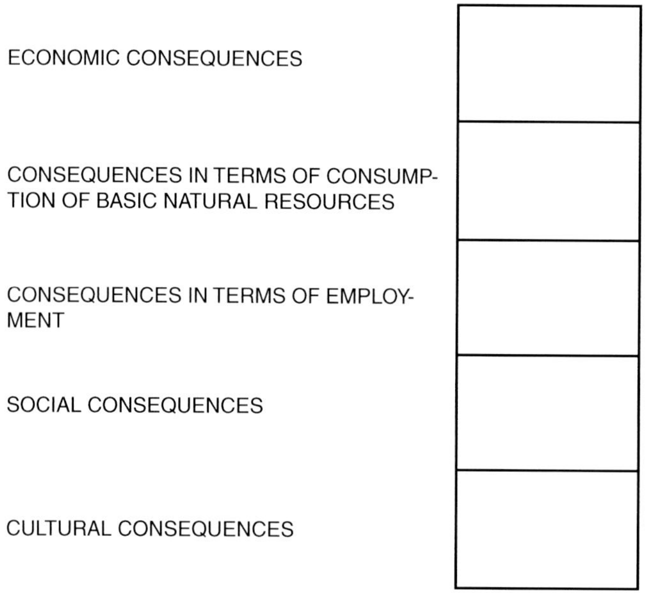

Fig. 2.3. The five equal areas of social consequences from tourism.

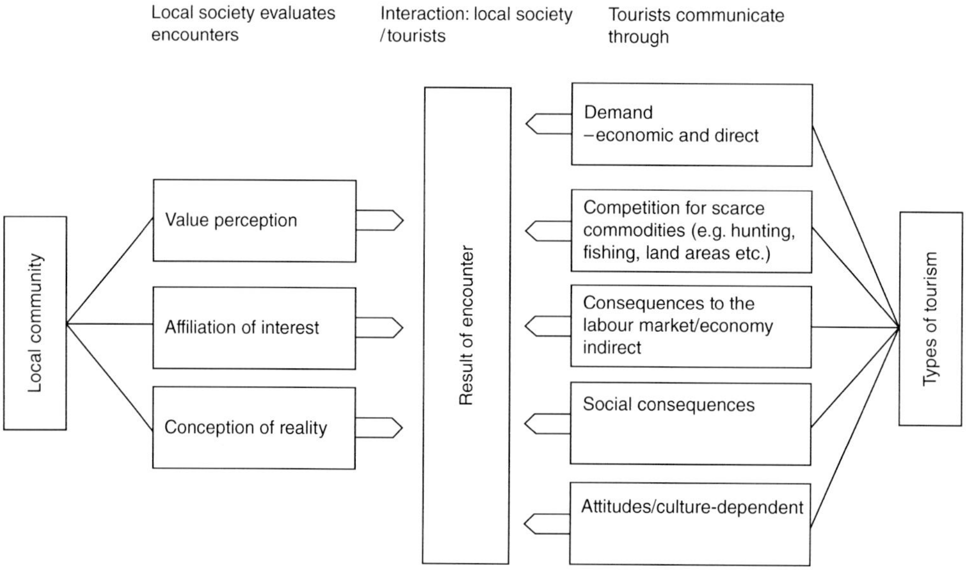

Fig. 2.4. The encounter between local society and visitors.

Notes

[1] I will use the terms *travel-business* and *tourism* synonymously. Tourism is a phenomenon that must be delimited in time and space. In space, a downward delimitation must be set, since movement must have a minimum scope. One becomes a traveller only after one has moved away from one's home place. One's home place is where one is on the census roll and pays taxes. The delimitation in time is vertical, since one's absence must not be so lengthy that the traveller becomes a resident of the new place. For the purposes of my analysis, it will suffice to distinguish a tourist from an immigrant (juridically speaking) and one who practises local recreation by saying that anyone who stays outside his/her home municipality, who has financed the trip from his/her home place and intends to return within a maximum of one year, is a tourist. This is a delimitation that conforms with the delimitation of terms in the major work done by English economist Ogilvie (1933, pp. 3–6). Ogilvie's work is considered an early pioneering achievement and a classic. Up until this work, the English term 'tourist' had been synonymous with 'a visitor on holiday', and 'travelling for pleasure'. Ogilvie makes the term more comprehensive so that our two terms – 'a traveller' and 'tourism' are such that I choose to use these corresponding terms. In his clarification of these terms, Ogilvie states that: 'For economic purpose, then, tourists are all persons who satisfy two conditions, that they are away from home for any period less than a year and, second, that while they are away they spend money in the place they visit without earning it there' (1933, pp. 5–6). Ogilvie's innovation is that he makes the term 'tourist' more encompassing ('...business visitors, government officials or trans-migrants, who no less are tourists in the generic sense' (1933, p. 5)), since he does not associate the term with 'holiday' and 'travelling for pleasure' but rather to a time-delimited movement in space, and posits the basis for the trip itself (economically related) as linked to one's home place.

[2] Flognfeldt Jr has added the following conceptual definitions to the figure:

> Local recreation is that recreation that can be conducted in conjunction with one's workplace or place of residence, i.e. when one is not away on a trip. Distant recreation is the recreation that is pursued while on a trip – frequently denoted by what we term tourism. Business-related tourism is the part of tourism that is not at the same time a part of recreation, that is, it is located outside the recreation circle. All types of outdoor recreation are at the same time recreation and they are therefore located inside the recreation circle. In some cases, one will have to travel to take part in outdoor recreation; in such cases, outdoor recreation is both a part of recreation and of tourism. In other cases, one may conduct outdoor recreation in conjunction with work or place of residence and thereby not participate in tourism.

[3] WTO stands for World Tourism Organization. WTO was founded in 1975 with the restructuring of the International Union of Official Travel Organizations (IUOTO, founded 1925) into an intergovernmental body. WTO has about 106 (as at 1984) member countries and is acknowledged by the UN as an international agency.

References

Aubert, V. (1969) *Det skjulte samfunn*. Pax, Oslo.

Baron, R.R.V. (1989) *Travel and Tourism Data: A Comprehensive Research Handbook on the World Travel Industry*. Euromonitor, London.

Boorstin, D.J. (1964) *The Image: A Guide to Pseudo-Events in America*. Harper, New York.

Cohen, E. (1974) Who is a tourist? A conceptual clarification. *The Sociological Review* 22(4), 527–555.

Faludi, A. (1978) *Essays on Planning Theory and Education*. Pergamon, Oxford, UK.

Flognfeldt Jr, T. (1976) *Reiselivets, rekreasjonens og friluftslivets plass i rekreasjonsprosessen*. Oppland Distrikthøgskole, Lillehammer, Norway.

MacCannell, D. (1973) Staged authenticity: arrangements of social space in tourist settings. *American Journal of Sociology* 79(3), 589–603.

Ogilvie, F.W. (1933) *The Tourist Movement: An Economic Study*. Staples, London.

Parker, S.R. (1965) Work and non-work in three occupations. *The Sociological Review* 13(1), 65–75.

Pearce, D. (1988) *Tourism Today: A Geographical Analysis*. Longman, Harlow, UK.

Ramsøy, O. (1978) Fritid og rekreasjon. In: Ramsøy, N.R. (ed.) *Det Norske Samfunn* (2nd edn). Gyldendal, Oslo.

Smith, V.L. (1977) *Hosts and Guests: The Anthropology of Tourism*. University of Pennsylvania Press, Philadelphia, Pennsylvania.

Turner, V. (1974) *The Ritual Process*. Penguin, Harmondsworth, UK.

Wagner, U. (1977) Out of time and place: mass tourism and charter trips. *Ethnos* 42(1–2), 38–52.

3 Localization of Tourism

Sondre Svalastog

Introduction

Conditions for localization change over time. Technical conditions change; demand for resources, skills and needs change, along with perceptions and symbolic values etc. Tourism consists of production exposed to competition and protected production. A business that is exposed to competition and at the same time resource-based has stringent localization requirements if it is to be competitive. Only the existence of the very best resources can provide an adequate basis for success. In this chapter, therefore, we will spend time considering localization and resource issues.

Factors in Localization and Production

In conjunction with localization and production, we use the terms *localization factors* or *localization determinants* and *production factors* or *input factors*. Any localization and/or production analysis must necessarily be rooted within these two dimensions.

Localization factors are the *total framework conditions* surrounding an economic activity, while the cluster of production factors is limited to the *specific input* in a production or an economic activity. The cluster of production factors is a key localization factor but, as such, it is treated as an aggregated quantity (cf. the section 'The Location' in Smith, 1981, pp. 45–46).

To preclude any misunderstandings in what follows, it may be wise to state more precisely the following three aspects. First, it is an established fact that all production and economic activity is dependent on a set of *input factors*. These may include unprocessed physical input factors, know-how, various product components and other intangible conditions of a cultural and social nature etc. Therefore, for purposes of pedagogy and conceptualization, a common technical terminology in the subjects of geography and social economics has been developed and applied to the localization factor that has had the greatest impact on localization. Site-specific resources are crucial. Production or economic activity will thus be denoted as *resource-based* or *commodity-based*. Production and economic activities will be termed *market-based* or *market-oriented*. Examples of this kind of activity in the service sector are doctors' offices, solicitors' offices and cinemas (Holte, 1982, pp. 104–108; Paterson, 1976, pp. 214–221). Correspondingly, one also has to have *energy*, *labour* and *network-based localizations* etc.

To characterize the main forms of localization of tourism means using a *system of concepts* that [due to its specificity] at the same time is informative and operational. The conceptual strategy is chosen to avoid a particular form of

DOI: 10.1079/9781800621480.0003

circular reasoning that occurs when tourist-related production is simply described as a production in need of input of resources and therefore is resource-based, which is a description that is no more than a tautology.

Second, we need to make ourselves conscious of the fact that we are dealing with two main categories of localization factors and two categories of production factors. The main category comprises *material things and conditions*, for example access to land area and energy, and climate. The other main category consists of *intangible factors*, for example competence within the context of social and cultural locations. Technological development so far has made the *intangible factors* increasingly important for localization (De Blij and Muller, 1988, p. 225): '... a new set of locational forces has been unleashed wherein the distribution of key economic activities is now being shaped by noneconomic variables ...'). The direct cause of this is that it has been less expensive to move important input factors such as commodities and energy. An additional, exacerbating factor has arisen as corporate management has become professionalized and bureaucratized (no longer affiliated with the company founder, but a team of various management professionals that Galbraith (1967), has called 'the technostructure'). This group of professional managers, marketers and product developers is geographically mobile because they have minor ownership interests (or none at all) in the companies they work for and they therefore place greater emphasis on environmental and cultural factors in their workplace and home community. These two factors generally work together and move towards central, urban areas with more advantages.

Third, our era demands that we recognize localization factors, which encourage us to establish a *dynamic localization analysis* rather than a *static* one. We need to include wage and cost levels in our local area as real, key localization factors and let the K variable (know-how, creativity, communication and culture) become the dynamic element (Andersson and Strömquist, 1988, p. 29). It is a static and short-term strategy to use wage and cost levels for 'primitive price competition and cost-hunting' (see, particularly, the section 'Dynamic competition', Andersson and Strőmquist, 1988, pp. 23–24). A high wage and price level is a localization factor that warrants great emphasis on knowledge enhancement aimed at mastering demand and product development, implementation of production processes, mastery of rapid changeovers and adequate interaction with a diversified market.

Main Forms of Tourism in a Localization Perspective

From the perspective of localization and integration, we have the following three main forms of tourism:

The main form is *market-based tourism*, encompassing business-related tourism, oriented towards densely populated areas where production must take place, and people on business-related journeys access their customers. In conjunction with localization theory, I therefore choose to call this part of tourism 'market-based'.

In terms of adaptation this part of tourism affects:

- economy and employment;
- land area;
- society;
- culture; and
- local politics/policy.

This segment of tourism is unproblematic, or, more aptly put, it is easy to integrate into the receiving areas. Visitors are able to pay (are a part of the sheltered part of the industry), demand is distributed evenly throughout the year and visitors interact with like-minded people, merging with the social control of residents. If we measure the volume in terms of overnight stays outside the tourists' home location, this main group accounts for 10% of the total travel volume in Norway (Svalastog, 1987, pp. 26–27, 53). The corresponding percentage for Sweden, according to Aronsson (1989, p. 34) is 8%.

The leisure sphere presents us with the other two main forms of tourism. We will term one of these *network-based tourism* or 'visiting tourism'. This group comprises mainly travellers visiting families, acquaintances and friends. The bulk of these live in private accommodation and comprise a significant portion of *non-commercial tourism*. They roughly reflect the geographic settlement pattern – but, nevertheless, with a certain preponderance of people travelling from urban to rural areas. This is largely due to our historical situation, where we have had scattered primary industry settlement and would therefore rather 'travel to the country' in our free time than to the city. Rural villages are where our roots are based, with which we want to remain in touch.

In terms of adaptation, this part of tourism entails consequences for:

- economy and employment;
- land area;
- society;
- culture; and
- local politics/policy.

This segment of tourism is also unproblematic to integrate locally. It comprises the bulk of the 'anti-tourists'. Network-based tourists in Norway account for about 45% of the total traveller volume. The corresponding percentage for Sweden, according to Aronsson, is 40%. The traffic rate mirrors the holidays and days off throughout the year and the segment generates relatively little economic activity. Demand is located in the protected part of tourism.

The last main group, and the one that represents the greatest challenges in terms of 'integration of tourism in local communities', is linked with what I have chosen to call *resource-based tourism*. This is the segment of tourism that is governed by the *recreational opportunities, activities and experiences* in a given area. (It is worth noting that one can justifiably equate the academic term *resource* with the colloquial term *attraction*).

This part of tourism is highly complicated and produces impacts on:

- economy and employment;
- land area;
- society;
- culture; and
- local politics/policy,

which will vary to a great extent. In a technical sense, this main tourism group will require much knowledge and understanding to enable it to achieve a proper integration and localization in terms of volume. The volume of sun-worshipper tourism is plotted at the lower end of the triangle in Fig. 2.2 (see Chapter 2). From a local community perspective, adjustments to this type of tourism are irreversible. Sun-worshipper tourism accounts for the remaining 45% of the volume. The corresponding percentage for Sweden, according to Aronsson, is 52%.

The main forms of tourism related to localization can be segmented and are shown in Fig. 3.1.

(A)

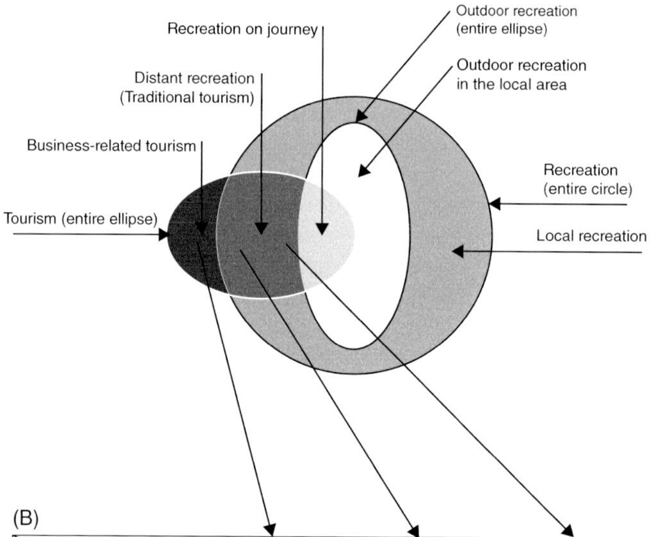

(B)

MAIN FORM: / VOLUME MEASURED IN GUEST NIGHTS	MARKET-BASED	RESOURCE-BASED	NETWORK-BASED
NORWAY	10%	45%	45%
SWEDEN	8%	52%	40%

Fig. 3.1. Tourism segmented into main forms for purposes of localization. (Part A adapted from Flognfeldt Jr, 1976, p. 16; Part B from Svalastog, 1987, p. 27)

* Note: The term 'recreation', found in the above diagram, is used in its universal sense, i.e. as a synonym for leisure/free time (added by S.S.).

About the Derivation of Conditions

To be able to say anything definitive and objective about the conditions an area must possess for resource-based tourism, one must first analyse the following six determinants. These comprise five factors linked with the production site itself (treated under Point A, in Table 3.1a) and one factor linked with the market (treated under Point B).

First determinant *The geographic location of the area.* Whether it is centrally located or peripheral in relation to the markets, i.e. where demand is present.

Second determinant *The social infrastructure.* Roads, railways, boats, ferries, airports etc. Water, sewerage and other social infrastructure.

Third determinant Clustered resources that it may be worthwhile to group under the following four main types:

- Natural resources, including climate, flora, fauna, topography, water, geology and extent/dimension of the area.
- Culturally or socially determined resources with subgroups. Material resources such as monumental buildings, installations, relics, prehistoric features etc. and intangible factors linked, for example, with the place's history as a source of culture. Cultural centres of attraction, or other social and environmental aspects that make the local community interesting for visiting tourists.
- Synthetically created resources, meaning invested capital in building up tourist attractions. Good examples from abroad are Disneyland and Legoland. Examples from Norway might be Lilleputthammer in Øyer municipality, Kongeparken at Jæren or Tusenfryd outside Oslo.
- Competence (or human resources) that comprise both expertise possessed by a few people and a pool of general knowledge within a natural recruiting area.

What is important to bear in mind for our purposes is that natural and culturally dependent resources are site-specific, whereas synthetically created resources and competence resources are not site-specific to the same extent. The latter two can be acquired over time through decisions.

Fourth determinant *Tourism-related infra- and suprastructure.* The tendency to commonly group tourism-related infrastructure and tourism-related suprastructure together is due mainly to the double role that accommodation and eating establishments play. They feature both attractions and infrastructural functions. The ski-lift moves in space, while the hotel bed moves in time – but with the bed comes the building's inner landscape. It cannot be separated in the same manner as the ski-lift from the landscape.

Fifth determinant *Wage and cost levels* are a heavy-handed segregation mechanism. Accessibility may be good; the social infrastructure well developed, abundant availability of desirable resources and the basis of tourism-related infrastructure and suprastructure may also be good. Despite all this, however, stakeholders in an expensive country increasingly find that it is not profitable to begin production of certain types of products because other areas can offer equivalent products at a lower price. In a schematic sense, we are dealing with the following relationships:

- Areas with low wage and cost levels produce standard goods in larger quantities.
- Areas with high wage and price levels (along, usually, with a higher level of knowledge) must develop more complex and competence-dependent products.

Sixth determinant *The market with its demand and payment capability.* An area that wants to establish what opportunities it may have for resource-based tourism must first take an inventory of its production-related assets and match these against the market (see Fig. 3.2 and Table 3.1a for a summary).

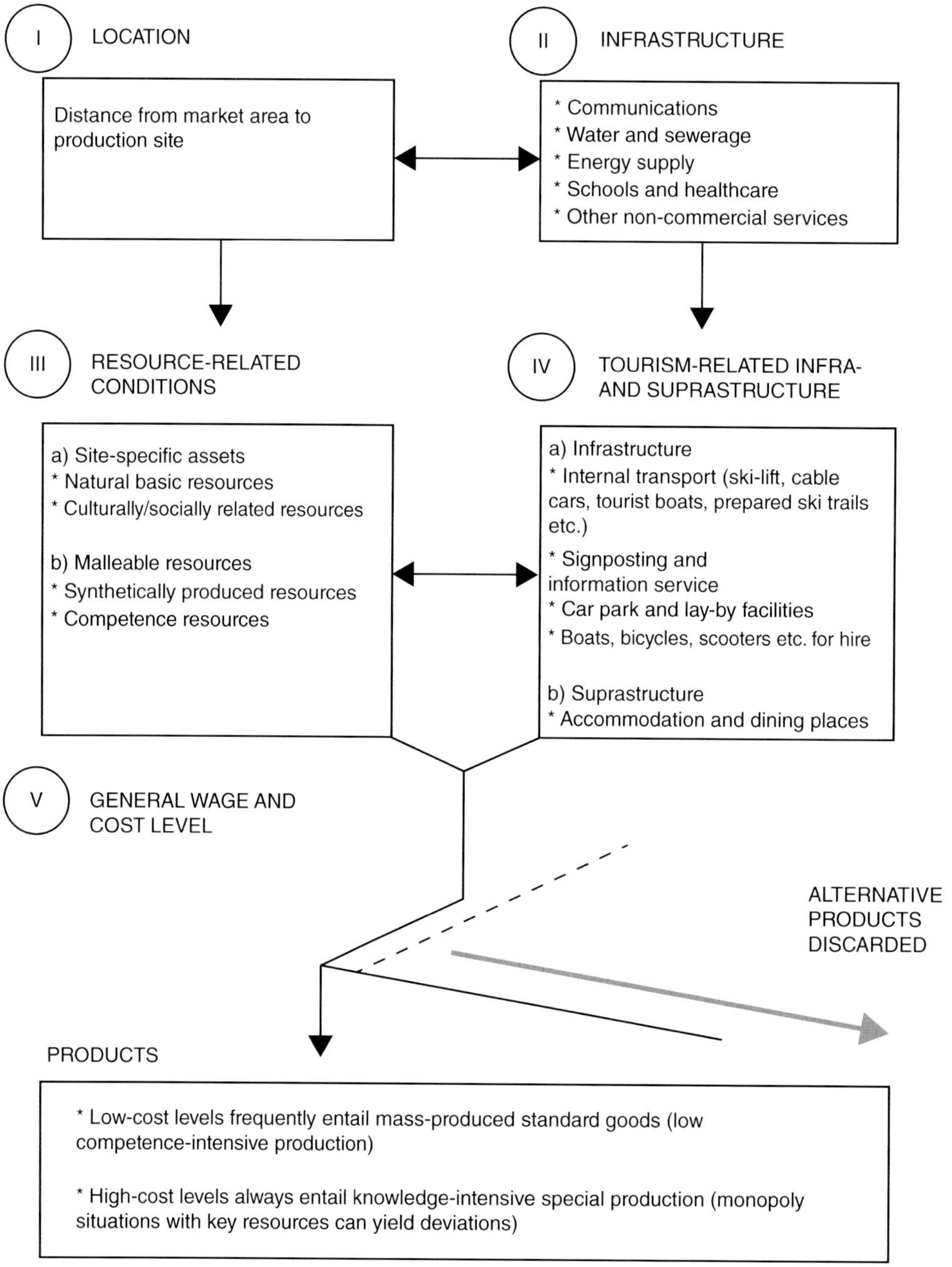

Fig. 3.2. Resource-based tourism and determinants of localization (factors that determine place, space and scope).

Table 3.1a. Resource elements from Fig. 3.1 (Box III) specified.

A. SITE-SPECIFIC RESOURCES

I. Basic nature-related resources:
- climate
- flora
- fauna
- oceans, water and watercourses
- geological conditions
- topographical conditions
- scope of areas

II. Culturally dependent resources:

(i) material things and conditions (concrete resources)
- prehistoric features
- monumental buildings and installations
- cultural landscape (how humans have adapted to the place's conditions for self-realization and production) etc.

(ii) intangible conditions (abstract/intangible conditions)
- cultural places of origin
- cultural producers of the current era
- cultural factors via social interaction, customs and practices etc.

B. MALLEABLE RESOURCES

III. Synthetically created resources (capital transformed into tourism's resources/ attractions):
- activity installations, play areas etc.
- zoological parks
- amusement parks of different types
- futuristic constructions, fantasy land etc.

IV. Knowledge resources:
- a few experts who are able to engage in product development and offer general advice
- the general level of knowledge (a decisive factor for the quality of the labour force)

Table 3.1b. Information of resources available at a location, including climate.

Guaranteed days with snow cover	Ensuring a place offers snow sports
Number of summer days	One of the key factors for sun and outdoor activities
Number of clear days	What is the probability of good conditions for outdoor activities?
Humidity	Does the place have a dry or humid climate?
Wind velocity	Significant for location of sports activities and recreational tourism
Length of days	In Norway, the midnight sun areas are particularly important

Important Aspects in the Phenomenon of Resources

Resources are relative, not static, entities, i.e. things that change over time. In his classic work *World Resources and Industries*, Zimmermann (1972, p. 8) writes:

> It is certain that materials and concrete things can function as resources, and they play a very great role as resources. Just think of ...
>
> Unless easily discernible factors such as knowledge, sound policy, health, social harmony, freedom etc. are ignored, despite the fact that these may possibly be more important than all the iron, coal, gold and silver in the world...
>
> Third, there has been a very unfortunate tendency to think of resources as separate things, such as coal, for example, rather than as an entire complex of materials, forces, properties, contexts, institutions, political measures etc. That each individually can help to

explain the way coal functions as a resource in a given place at a given time...

This also applies for tourism resources and I shall use Lillehammer and the time span 1854–1990 as a limited period to demonstrate how conditions are changed by:

- techniques;
- values;
- skills;
- consequences of changes outside the area (exogenous changes);
- consequences of changes within the area in question (endogenous changes);
- sociological (relative) changes between competing tourist locations; and
- wage and cost changes between countries.

The following assumption is made at the beginning of the analysis: Lillehammer's leisure resources were at their most depleted at the end of the 1980s, as resources for tourism had changed but Lillehammer had not kept up with these developments. Furthermore, important resources had been decimated during this period and technological changes in Norwegian society had further weakened Lillehammer as a tourism area. For concision, I focus on a key number of subjects:

1854 The Oslo-Eidsvoll railway was opened when there was already a boat connection across the River Mjøsa (Eidsvoll-Lillehammer). This meant that the speed of the trip from Oslo to Eidsvoll was practically quadrupled (from about 7 km/h to about 25 km/h, see Table 3.2). The entire stretch from Oslo to Lillehammer was upgraded to ensure travel comfort which no other inland areas could compete with. In terms of communications, Lillehammer had no competition up until 1892, when the Bandak–Nordsjø canal was completed and the Bergen line was opened in 1909. Indeed, in 1854 Lillehammer was granted a social infrastructure advantage that it alone enjoyed until 1909. (See Table 3.2 for more detailed information on the advances of communication, and Figs 3.3–3.5 for more about the individual stages.)

1880–1950 This is the period of the 'Lillehammer colony' of painters who lived and worked here during great portions of the year. Their emergence was a result of four contributing factors. First, good communications to Oslo (social infrastructure); second, the availability of various hotels and boarding houses (tourism suprastructure); third, low price levels compared with Oslo; and fourth (the most decisive factor), the air quality and beauty of the natural landscape. The interplay between these factors is well documented by conservator Per Boym (1990; see in particular pp. 5, 41–42).
Currently, Lillehammer is Norway's fourth most polluted city from nitrogen dioxide. Only Stavanger, Drammen and Oslo have higher levels of NO_2.
Lillehammer greatly exceeds the WHO's critical limit for nitrogen oxides (Sjøsten, 1990, p. 10).
This is an example of how external changes in production can weaken the resource potential (clean air) in an area. Central Europe has many similar cases (often in combination with air and noise pollution) that have ruined earlier conditions favourable to 'spa resort tourism'.

Table 3.2. The great advances in communications. (Partly from Bjørnland, 1977)

Means of transport	Speed	Speed in relation to basis
Horse and sledge	3–4 km/h	basis = 1
Horse and wagon	5–4 km/h	about 1.25
Steamboat about 1840	12–15 km/h	about 4
Railway 1854	25–30 km/h	about 7
Private car about 1930	*c*.40 km/h	about 10
Private car about 1960	*c*.60 km/h	about 15
Aeroplane about 1980	*c*.750 km/h	about 200

1870–1970 Changes in values of holiday travellers. High season was originally associated with the off-snow period and the main activities included socializing in the salons and hiking (on foot during summer, on skis in winter). Lillehammer is surrounded by ideal terrain for these activities.

From the mid-1960s and for the next 20 years, tour skiing gave way to a preference for alpine skiing as Lillehammer lacks the topographical features necessary to develop a comprehensive and competitive alpine skiing product.

This change of preference in the market explains why a systematic expansion was started in the mid-1960s, for example in the areas of Hovden in Bykle municipality in Aust-Agder county, Hemsedal municipality in Buskerud county, Beitostølen in Øystre Slidre Oppland county and Trysil municipality in Hedmark county, which has been ongoing up until now.

1945 As mentioned earlier, resources are somewhat relative, and it is easy to make wrong decisions if this is not kept in mind.

A comparison of the winter sports resort towns of Geilo and Lillehammer throughout the post-war period illustrates this. Lillehammer, which was a railway stop during the first years of the period, enjoyed asphalt streets, pavements, street lights, parks and benches, short distances to tourist resorts, art collections, cinemas, museums etc. In stark contrast, Geilo was the complete opposite. The streets were filled with mud puddles, there were no pavements, the distance between resorts was relatively far and there were no galleries, swimming halls, souvenir shops, bookshops etc.

Over the course of the next 40 years, this would change and Geilo began to develop the same amenities as Lillehammer, although with massive and detrimental car traffic, noise and air pollution.

Despite the fact that Lillehammer seemed independent, it had not changed at all during the period, which, compared to other locations, had far-reaching consequences.

1945 Wage and price development during the course of the post-war period catapulted Norway out of the low-cost range of countries and into the 'top ten' group in terms of salaries and costs. This would require reorganization. Products that could be produced at competitive prices in a low-cost situation could no longer be successfully produced. New products demanded a different kind of competence as well as a somewhat different production apparatus. Lillehammer met the new situation of the tourism industry by taking a defensive stance. Companies met the challenge by taking on advantages from economy of scale for its products.

1960 Changes in competence (e.g. foreign languages and swimming). Economic and technological advances set the scene for the tremendous growth in charter travel abroad. In a world context, Norwegians have become financially rich, visiting exotic places as 'social free ports'. Vacations have become holidays for large groups of people escaping aspects of social control. These are changes that do not selectively affect areas and thus do not affect Lillehammer to any specific degree. These vacation changes are felt by all Norwegian producers in this competitive part of the market.

1960–1980 The destruction of environmental resources such as low-lying, contiguous and characteristic buildings, replacing them with incongruous architecture; for example, replacing large industrial buildings in the downtown areas with the cheapest form of grey concrete block houses etc. In general, this was an environmentally destructive practice, so conspicuous that it became the object of discussion in the architectural periodical *Byggekunst* (Sveen, 1977, pp. 65–66), the local annal of Fåberg and Lillehammer (Sveen, 1980, pp. 116–127), *Aftenposten*'s morning edition (Kolloen and Larsen, 1988) etc.

With this incursion came destruction of resources from within by forerunners of the 'yuppie' culture who were given the opportunity to wreak havoc on the area.

1972 The next beneficial event that seems to have occurred in Lillehammer during the entire period between 1854 and 1990 is linked with the development of professional competence in the fields of teaching, research and consultancy services. The two-year tourism study programme established in 1972 at Oppland University College in Lillehammer was expanded in 1989 and became a three-year study programme. Furthermore, the Eastern Norway Research Institute established a Lillehammer department, with travel and tourism scholars in 1986. The most recent addition to the Lillehammer educational environment is the Norwegian Institute for Nature Research (NINA), which was established in Lillehammer in 1989. Together, these institutions, their subject areas and research assets represent, quantitatively, the most comprehensive centre of importance in the field of tourism in Norway. However, the point here – one that is important and at the same time easy to overlook – is that all three institutions function as national bodies and do not necessarily yield any advantage in particular for Lillehammer. Lillehammer itself must be organizationally capable of exploiting benefits but, so far, has not proved proficient in terms of such organization. My observations throughout the period 1972–1990 suggest that other municipalities and counties have been more successful in reaping the benefits (competency) from the Lillehammer environment.

The 1994 Olympics Change the Conditions

The conditions for both resource and market-based tourism in Lillehammer after the 1994 Olympic Games differed from those that existed prior to their assignment to Lillehammer (in Seoul on 15 September 1988). So what important types of conditions were changed? *A priori*, one must assume that the situation before and after will not change any of the natural conditions. The elements of climate such as precipitation, temperature, number of clear days, wind etc. will not affect the characteristics of the area, at least in terms of the event itself and the investments made in it. The topography is likewise unchanged as is the flora and fauna. Moreover, the physical location of Lillehammer remains the same. The changes, then, do not lie with natural resources but with the human resource base.

If we consult Fig. 3.2, we see that the 1994 Olympics do not entail any changes in terms of Box I, 'Area's geographic location'. We can also discount changes in relation to Box V, 'General wage and cost level', but something significant will very likely occur in terms of what the area is capable of producing in the form of special products. We shall comment on this later. What remains, however, is Box II, 'Social infrastructure', Box III, 'Resource-related conditions' and Box IV, 'Tourism's infra- and suprastructure', and it is precisely within these areas that the conditions change.

Intuitively, we see that the 1994 Olympic Games entail significant changes in the social infrastructure and the tourism-related suprastructure. Communications have been significantly upgraded. This pertains to the road network between Oslo and the Olympics region as well as to the railway. In a purely physical sense, the location has remained the same prior to the decision to grant Lillehammer the Winter Olympics, but shorter travel time and more comfortable transportation have improved market access. The second and possibly more substantial change has taken place within the school, education and research sphere. In a national perspective and as a result of the 1994 Winter Games, Lillehammer will emerge as a national centre of importance in these activities. Firstly, this will result in an increased volume of profession-related traffic (market-based tourism), but of course also as a basis for increased resource-based tourism. The area may gain increased attraction value by virtue of this educational development.

The infrastructure and suprastructure of travel and tourism will be quite different following the 1994 Olympic Games compared with Lillehammer before the Olympics. This is necessary to be able to accommodate increased traffic to the area, but it does not generate traffic in and of itself. It is mainly changes in Box II and Box III that generate more traffic as a consequence.

How then do the 1994 Olympics change Box III? What does not change are the individual components on the natural side of the equation.

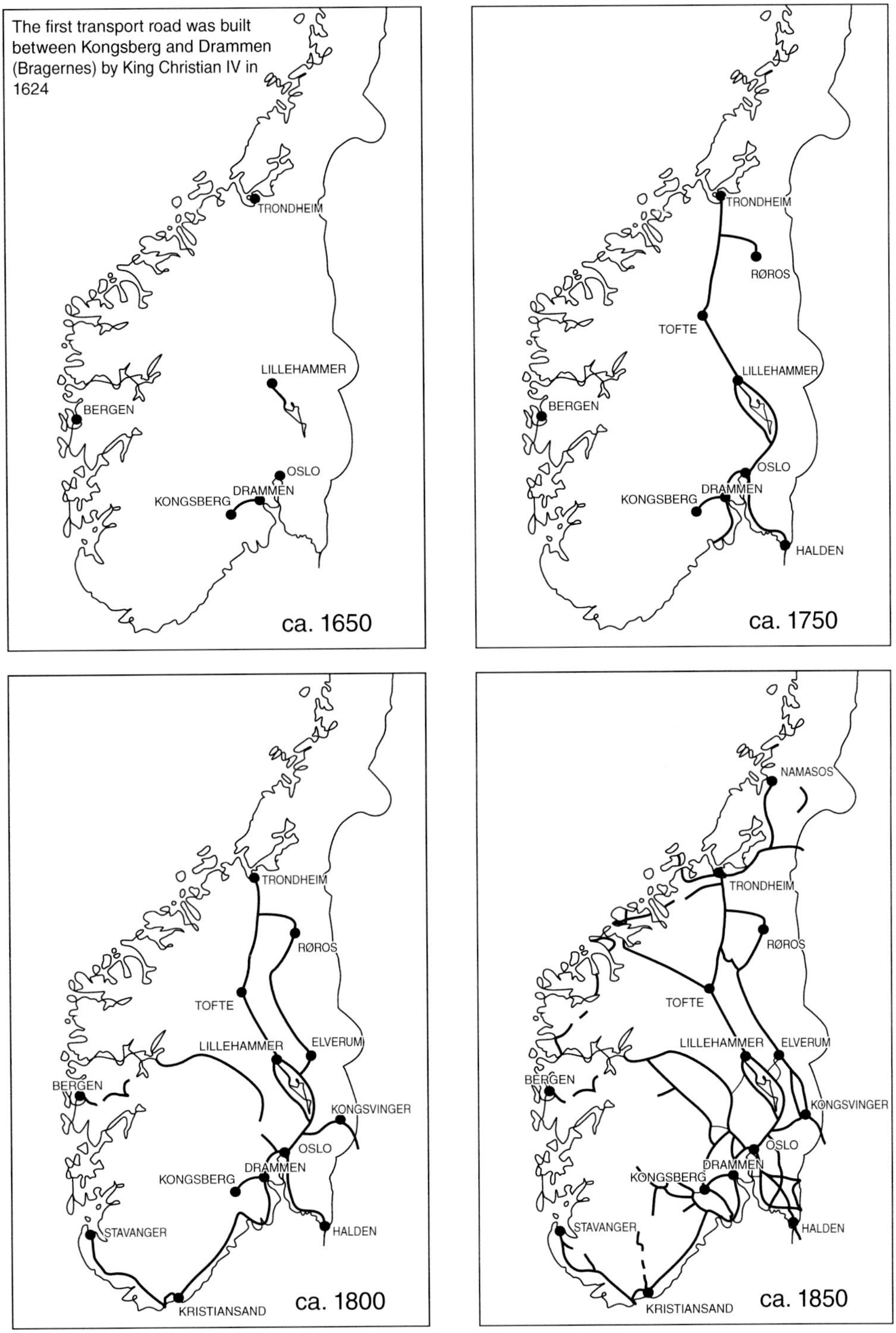

Fig. 3.3. Development of the road network for wheeled horsedrawn vehicles; transport speed of 6–10 km/h. (Adapted from Bjørnland, 1977)

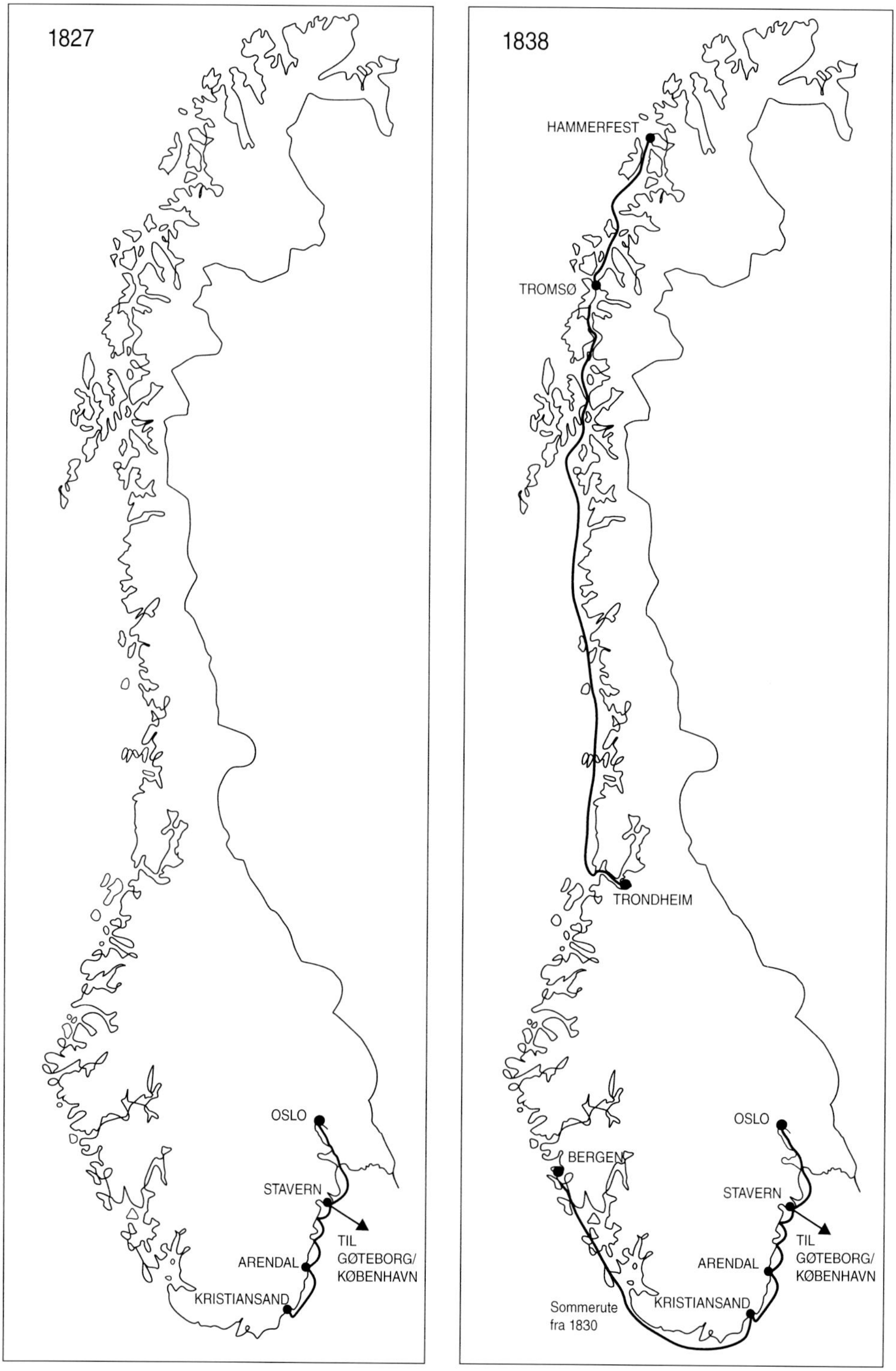

Fig. 3.4a. Development of the steamship routes; transport speed of 12–15 km/h. (Adapted from Bjørnland, 1977)

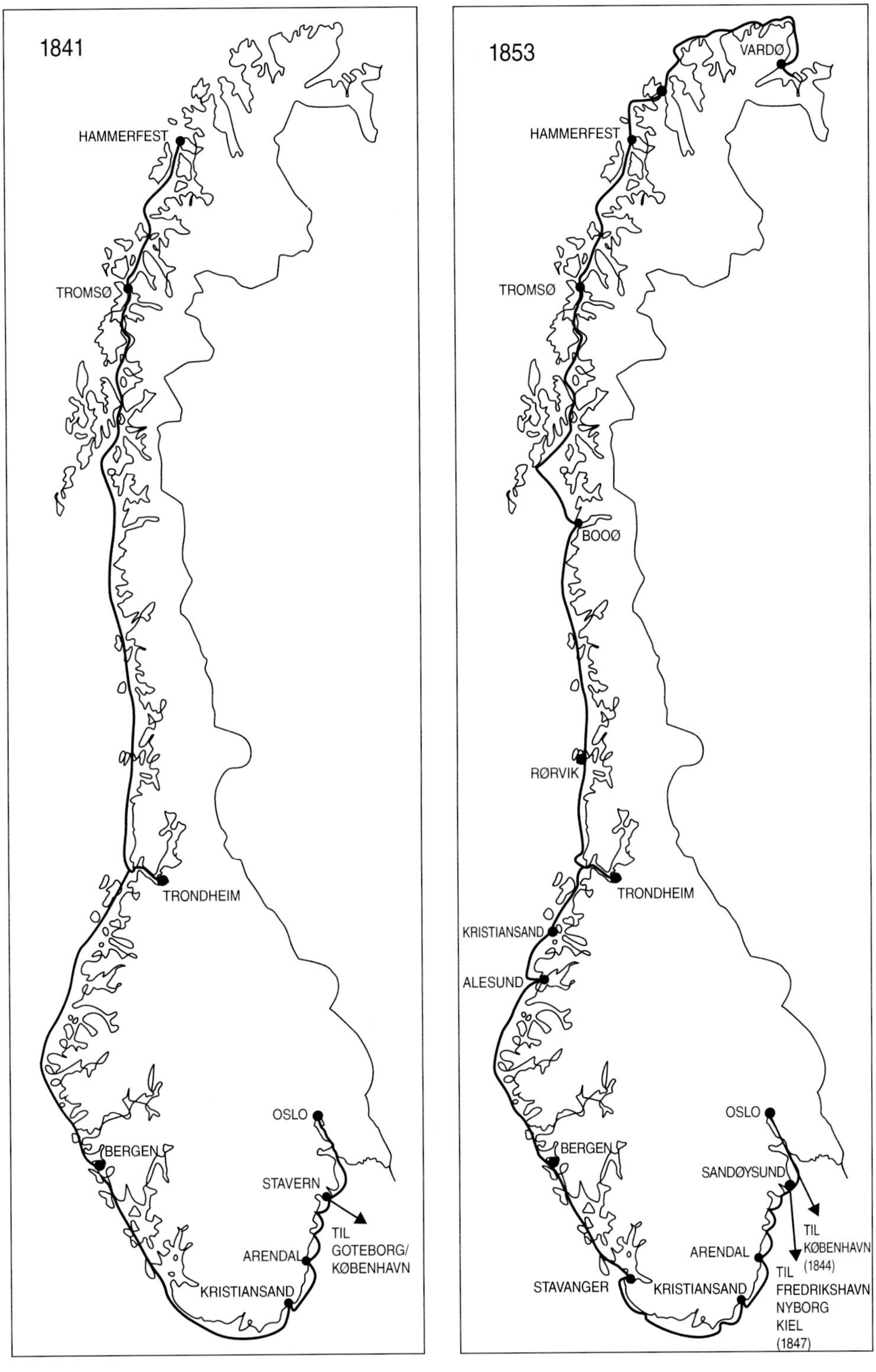

Fig. 3.4b. Development of the steamship routes; transport speed of 12–15 km/h. (Adapted from Bjørnland, 1977)

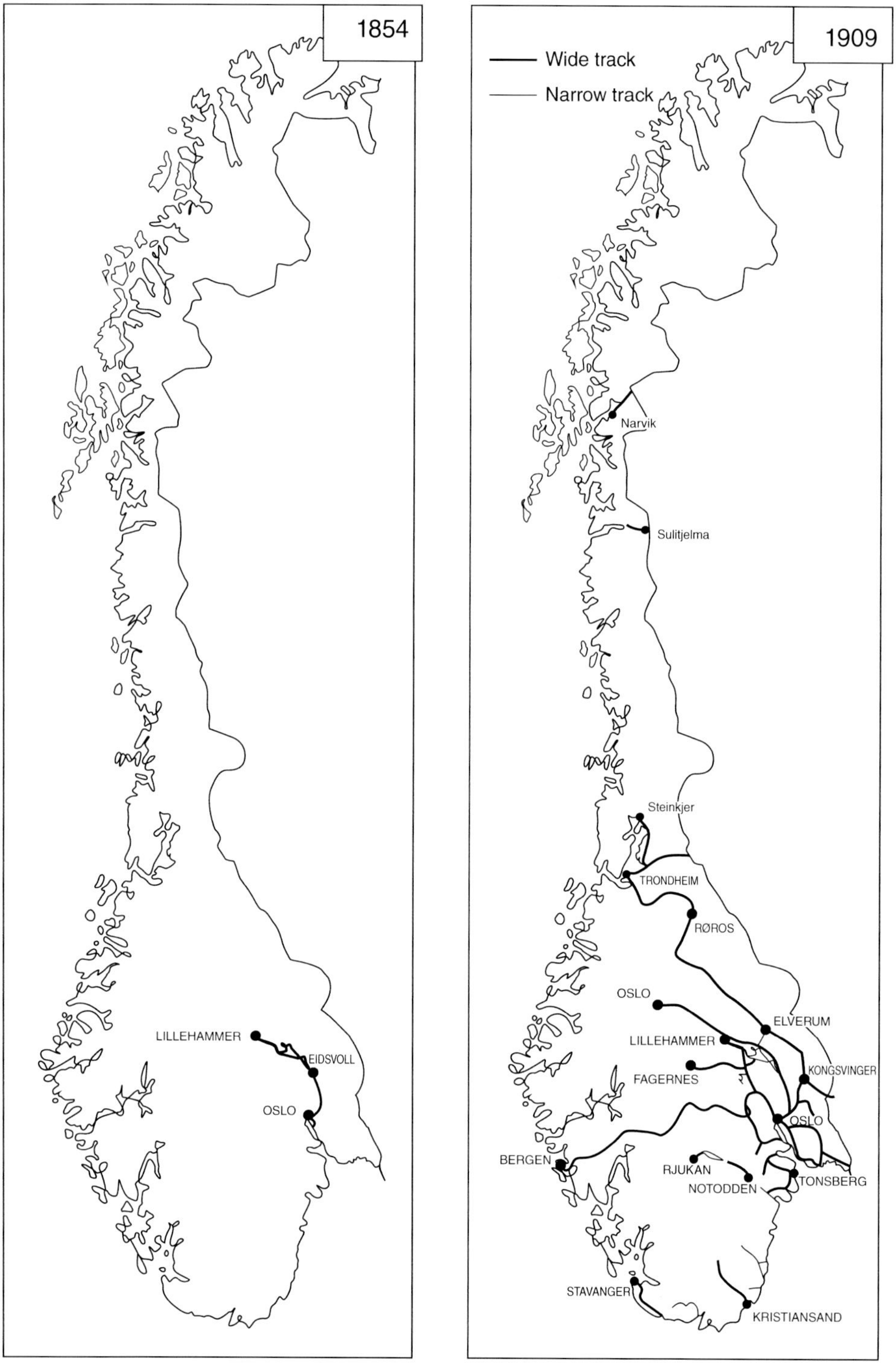

Fig. 3.5a. Development of the railway, 1854–1962; speed in initial phase 25 km/h. (Adapted from Bjørnland, 1977 and Østvedt, 1954).

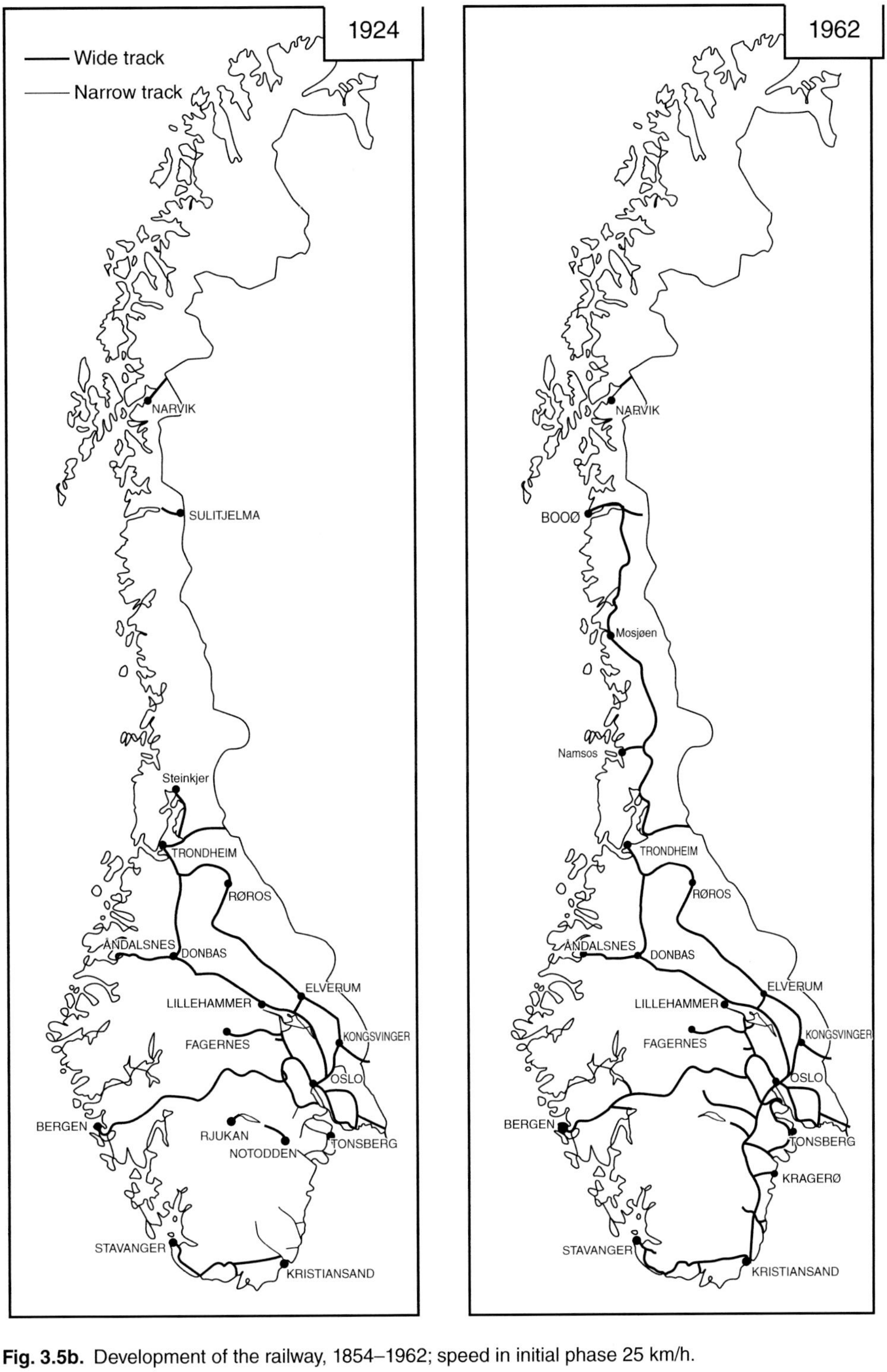

Fig. 3.5b. Development of the railway, 1854–1962; speed in initial phase 25 km/h.

Things change drastically, however, when it comes to the other three resource groups. The most conspicuous changes in resources within the main group of society-determined resources (attractions) are the following aspects.

Culturally dependent resources – material

The Sandvig Exhibitions (Maihaugen), which during the period 1 May–31 August 1992 had 147,000 visitors, i.e. Norway's 19th most visited attraction with a regular count, will be a totally different museum after expansion of indoor capacity by more than 70%.

The Lillehammer City Collection of Paintings, which already enjoyed a high status because of its collection of Norwegian art, increased its indoor area by more than four times as a direct result of the 1994 Olympics and increased its staffing from 3 man-years to 8.5 man-years in 1993. The collection will most likely be assigned the national status of a core museum for eastern Norway in the course of one or two years, and the increase in staff that will result from this status is budgeted at 2.5 man-years.

The Olympics Information Centre and Business Centre had a total of 280,000 visitors in 1992 (the corresponding figure for the period 1 May–31 August 1992 was 146,000 persons, which ranks the facility just below Maihaugen). The number of visitors for the first quarter of 1993 was 70,000 and represents an increase of some 70% over the same quarter in 1992. The centre's prognosis for the entire year (1993) is an increase to 560,000 visitors, which amounts to a doubling of its number of visitors.

Several of the new sports arenas will be considered attractions. The best example is, without doubt, the new ski-jump arena that was deemed to be aesthetically very successful and has a jump tower that provides an excellent introductory viewing platform for a panoramic view of Lillehammer and of the settlements along Lake Mjøsa. Experience so far suggests that the ski-jumping facility can attain roughly the same importance as the Holmenkollen ski-jumping hill in Oslo.

Culturally dependent resources – non-material

An equally important change in the area's culturally dependent potential may be found in the 'non-material' sphere. There is every reason to anticipate that Lillehammer as a destination for visitors will have a different image in the market after the Winter Olympics in 1994 compared with the time before the city was awarded the games.

Firstly, the Lillehammer area will have an image in the market as a place for winter sports and winter sport activities. Secondly, Lillehammer's image as a cultural city of national scope will be able to win back much of its erstwhile radiance, perhaps even with a somewhat broader scope. The portion comprising museums will be significantly bolstered, relatively speaking. The other image change on the cultural level will be linked with Lillehammer as a centre of educational competence. This dimension is significant because the museum aspect largely represents a bygone time, whereas education and its projects look to the future. Furthermore, these changes can be seen as supplementary and supportive of each other.

The Winter Olympics in 1994 have not necessarily brought in any synthetically created resources, at least not under the concepts that I use. However, the Olympics will provide the Lillehammer area with some very interesting expertise in terms of special products within resource-based tourism. Part of this will be connected to staffing of many of the sports arenas and an amount of skilled workforce that Lillehammer will receive when the Norwegian Upper Secondary School for Skiing establishes a department here (the main centre for the winter sports).

In summation this section illustrates that:

- resources are somewhat changeable in a functional sense, although a large number of physical resource elements remain constant;
- natural resources, including those related to tourism, amount to only a limited portion of the resources; and
- Boxes II, III and IV (Fig. 3.2) will be able to include and systematize for analysis all the changes that a given area experiences, either

because of changes to the product through time or the impact these changes have on conditions; for example, as seen above, because specific events change the conditions in a very short period of time.

Specifics for Resource Analyses

It is asserted that we have advanced to the point that we acknowledge that a significant part of tourism is resource-based and that the resource aspect consists partly of material factors and partly of non-material factors. The foundation is then laid for great academic progress.

Firstly, progress has been made because the improved concept, as demonstrated above, enables us to work more purposefully with the challenges of tourism development; the experience is actually akin to removing a blindfold from the eyes of a person who is about to move forward. Secondly, we can bring in experiences from other resources where resource mapping has a long tradition.

Personally, I like to draw parallels with mineral surveying. If we consult the mineral surveys for Norway, we find that all the municipalities in the country have one or more registers of mineral deposits. Furthermore, we become aware of the following three circumstances:

- In terms of minerals, there are large numbers of mineral deposits. The same is true of the field of resources for leisure-related travel and tourism.
- Only a small number of known mineral deposits provide the basis for profitable operations today. In other words, only the very best mineral deposits enable one to launch financially profitable operations.
- Resource mapping is also a description of non-political activity; one would not think so if one observes how politicians and bureaucrats involve themselves in the selection of geographic areas of leisure-related tourism (cf. with the work the Geological Survey of Norway performs). Secondly, under this point: surveying and describing resources for leisure-related tourism should be a continual ongoing task in the same way as mineral surveying is a continually ongoing endeavour. Among other factors, this is because:
    - o resources change over time in an objective sense;
    - o resources change over time in a relative sense;
    - o values and skills on the demand side change over time and thereby formerly neutral materials become resources and *vice versa*;
    - o accessibility of resource deposits change over time due to changes in technology and society's infrastructure; and
    - o new knowledge on the production side creates the need for other resources etc.

Resources and the Market

Tourism and travel marketers often speak disparagingly about product developers of today as being too product-oriented and insufficiently market-oriented. Producers need to become more market-oriented in the conventional sense of the sector.

A starting premise for the project was 'integration of tourism in local communities', which is not an easy approach to the challenge of developing tourism products. Consequently, it is suggested that a logical outcome for this approach has to be:

- resource mapping to determine which products can be developed;
- ascertaining from the market which products might be financially successful and then, as a general rule, a substantial discarding of products that fail to meet market viability;
- testing clusters of products that may have an economic market basis;
- testing the local community where the production is to take place – i.e. allowing local culture and local natural resources to be the decisive factors of production and scope of production.

Norwegian tourism is characterized by sparse product development during the past 25 years. The sector has met the future by:

- applying economies of scale; for example, the Norwegian Regional Development Fund has been an active helper (Svalastog, 1992, pp. 109–116; see, in particular, the section 'The rationalization period'); and

- increasing marketing efforts for ordinary (old) standard products; NORTRA, Norwegian tourism's most important marketing apparatus, has mainly prioritized its resources for this one task.

The result has been a continually diminishing competitiveness of the Norwegian tourism product. During the entire 25-year period, while we experienced a steady global rise in travel volume, the Norwegian portion of the market went into stagnation from the mid-1980s. The last decade has shown an absolute decline in profits in the leisure-related market. The balance of trade for tourism has been increasingly negative, even if we disregard a temporary deviation during the period 1987–1990. In the traditional rural form of tourism involving rental of cabins, we have experienced a decline over a period of several years in both rental time and the number of units the property owners make available. (On the subject of resource-based tourism in Norway, see Magnus, 1987, p. 7; NLVF study report no. 146, 1988, pp. 21–23; and Svalastog, 1988, pp. 103–120).

Know-how for Product Development

The traditional approach for visualizing a basis for product development has partly been to ask purchasers of products (by visiting the area where production is carried out) about their product satisfaction, supplemented on occasion by completing surveys into what are considered to be the main markets.

Market knowledge of this type can be useful:

- for improvement of ongoing production; and
- for measuring interest in the market for known types of products not produced in the area (to ascertain whether to begin replication of production).

The demand side can be surveyed through various interview methods, but it yields only external knowledge about the market. Latent demand cannot be revealed through interviewing, but only exposed by working professionally with local producers to understand demand from the inside – by determining which needs, motives and driving forces exist to enable tourism producers to form new product developments. It is by understanding demand from the inside that one can establish the basis for product oscillations. This is why a systematic development of knowledge through study programmes such as tourism sociology, tourism geography, consumer theory, history, ethnography etc. are so important.

To illustrate this point, let us consider a couple of examples: What would have happened (it was also tried, partially) if the computer industry had asked consumers in the 1950s and 1960s to suggest ideas for product development? Or what would Phillips, Netherlands, have received in terms of consumer support for the development of laser technology for various sound instruments? The only thing target consumers might have been able to do was provide feedback on ways to improve sound fidelity, but it would have been unnecessary to organize a comprehensive market survey to arrive at so obvious a conclusion.

Summary and Conclusion

In summary, the following factors can be identified in aiding the development of tourism at a location, as shown in Table 3.3.

Resource perspective: production that places requirements on the resources available. In quick summary, the situation is the following:

- For daily leisure activities (after-hours school or work), one must utilize the conditions (resources) that are found in the local environment.
- For day outings and weekends, dependence is on resources available ('intermediate resources') because the activity radius of the visitor provides certain, but limited, choice options.
- For holiday-related demand, the basic resources must be outstanding ('outstanding resources').

This chapter has focused on localization theory in general and, specifically, the factors that guide the localization of tourism. Moreover, attention has been given to how time alters how localization and resources are assessed. The next chapter goes further and concentrates on what I have chosen to call the sociological method, which will explain the resource-related conditions for tourism.

Table 3.3. Propensity for localization of various types of tourism.

The main distinction between forms of tourism on the basis of the propensity for localization	Segmented for localization analysis (dividing the total market into parts on the basis of localization)	Localized to:
I. Market-based tourism	a) Individual journeys that are professional, service- and business-related	Cities, industrial sites and communication junctions
	b) Locality-dependent	Where people live
	c) Demand from people travelling	Traffic junctions and through-arteries
	d) Trips to obtain	Cities and densely populated areas
II. Resource-based tourism	a) Holiday trips	Where resource conditions are found (natural, cultural, human-created)
	b) Brief leisure-related	Where resource conditions are found (natural, cultural, human-created)
III. Network-based tourism	a) Holiday trips	Where friends and relatives live

References

Andersson, Å.E. and Strömquist, U. (1988) *K-samhällets framtid*. Prisma, Stockholm.

Aronsson, L. (1989) *Turism och lokal utveckling: En turism-geografisk studie*. Kulturgeografiska institutionen, Göteborg, Sweden.

Bjørnland, D. (ed.) (1977) *Innenlands samferdsel i Norge siden 1800*. Grøndal & Søn, Oslo.

Boym, P. (1990) *Thorvald Erichsen på Lillehammer*. Lillehammer *bys malerisamling*.

De Blij, H.J. and Muller, P.O. (1988) *Geography. Regions and Concepts*. John Wiley & Sons, New York.

Galbraith, J.K. (1967) *The New Industrial State*. HoughtonMifflin, Boston, Massachusetts.

Holte, F. (1982) *Sosialekonorni*. Universitetsforlaget, Oslo.

Kolloen, I.S. and Larsen, G.T. (1988) En skam å vise verden! *Aftenposten* 30.11.88, Oslo.

Magnus, S.E. (1987) Vi må være pågående og aggressive i vårt samarbeid med myndighetene. *Norsk Hotell-og Restaurantblad* 1987(1), 7.

NLVF (1988) Om behovet for forskning og utvikling innenfor bygdeturisme. Report 146. Norges Landbruksvitenskaplige Forskningsråd, Oslo.

Østvedt, E. (1954) *De Norske Jembaners Historie*. Bind I–III. Norges Statsbaner, Oslo.

Paterson, J.H. (1976) *Land, Work and Resources: An Introduction to Economic Geography*, 2nd edn. Edward Arnold, London.

Sjøsten, B.A. (1990) Lillehammer lufta bedre. Gudbrandsdølen/*Lillehammer Tilskuer*, 30 January.

Smith, D.M. (1981) *Industrial Location: An Economic Geographical Analysis*, 2nd edn. John Wiley & Sons, New York.

Svalastog, S. (1987) Reiselivsressursene pa bygdene kan utnyttes bedre. *Norsk Landbruk* 12.

Svalastog, S. (1988) Tourism in Norway's rural mountain districts twenty-five years after the Mountain Planning Team's report. *Norsk Geografisk Tidsskrift* 42(2–3), 103–120.

Svalastog, S. (1992) Tourism in a changing Norway. *Norsk Geografisk Tidsskrift* 46(2), 109–116.

Sveen, W. (1977) Malurt i festbegeret. *Byggekunst* 3, 65.

Sveen, W. (1980) Lillehammers bymiljø: Små endringer, stor skade. In: *Fåberg og Lillehammer*. Fåberg historielag, Lillehammer, Norway.

Zimmermann, E.W. (1972) *World Resources and Industries*. Harper & Row, New York.

4 About Resource Analyses

Sondre Svalastog

Introduction

The recreational dimension (from the original Latin meaning: recreate = to create anew) is the tourist's prevailing motivation for taking a leisure-related journey. A carefully elaborated tourist typology, therefore, is the logical result of a leisure-related journey deriving from very different *driving forces, motives and needs.* Local recreation and leisure-related travel (including tourism as a function) are two essentially different phenomena. *The most important thing they share in common is that they both belong to the leisure sphere.* This difference, however, is not reflected in terms of models, hypotheses and theories of these two fields; in my opinion, this must be the most important explanation for why there has been so little academic progress over recent decades in the work involving analyses of resources available for leisure-related tourism. An academic discipline is based on the foundations of its concepts, models and theories. When those tools fail to provide suitable solutions to current challenges, attempts to present explanations and understand contexts will turn out to be wrong also.

The quest for understanding resources related to leisure tourism demands its own methods and models. First, this method must be capable of putting the researcher on the track to identify appropriate input factors/resources that provide environments, experiences and social conditions that appeal to the leisure traveller's demands. Second, the method must be one that can be used to evaluate and describe the resources and resource qualities. Third, the method must be one that provides expert assistance for:

- product improvements (this is necessary because we do not know if the products that are currently being offered in the market are optimal or first, second, third, fourth etc. ranked substitutions for something that could have satisfied the *effectuated demand* in a better way;
- promoting products and combinations of products that can meet the *latent demand*; and
- promoting products and combinations of products that can stimulate the *latent demand.*

Appraisal of resources (conditions for local recreation and tourism) has traditionally been considered a part of the geographical discipline, but not necessarily an effective one. Geographical methods have been shown to be academically applicable in the study of local recreational resources, but the geographical methods have not been equally successful for the study of resources for leisure-related travel and tourism. My working hypothesis is the same as that of the British geographer Brian S. Duffield, who concluded that 'the solutions of this intractable problem lie outside the geographical discipline' (Duffield, 1984, pp. 29–30).[1]

This hypothesis, that the solution must lie elsewhere than in geography as a discipline, is

DOI: 10.1079/9781800621480.0004

acceptable and in agreement with the approach I have chosen for this project, since resources are something that occur and exist in relation to human needs, motives and driving forces. The disciplines that lend themselves to an understanding of this, as I see it, are primarily sociology, social anthropology etc., not primarily geographical, or the discipline of landscape architecture. This will be my point of departure for what follows.

General Experiences with Resource Issues

Resource analyses and transfer of resource information to maps, for example, has a long tradition (cf. resource maps showing occurrences of industrial minerals, ores and natural stones, surface area maps for resources, climate maps, pollution maps, water resource maps, vegetation maps, landscape/topographical maps etc.).

In principle, resource analyses mean collecting and systematizing information on closely defined properties and factors that are of importance to a given purpose. In our case, the purpose is leisure-related (resource-based) tourism.

Universally, we know that:

1. Resources of interest usually occur:
   - scattered
   - with varying quality and richness/abundance.
2. Only the very best occurrences of resources provide the basis for profitable operations (competitive-edge products). Norway has a large number of registered mineral deposits but, despite this, only a limited number of mines in operation.

If we maintain the assertion that certain parts of travel and tourism are resource-based (anyone claiming otherwise will have to refute an immense rank of authorities (e.g. Clawson *et al.*, 1960; Coppock and Duffield, 1975; Chubb and Chubb, 1981; Travis, 1982)), it will be important to determine what the crucial resource conditions are (Travis, 1982, p. 257.) and how these can be discovered, measured and organized from survey information.

The analytical method we develop is entirely determined by our academic perception of reality. Therefore, let me repeat the main points of this and at the same time elaborate on a few of the details. In my academic perception of reality, I maintain that local recreation and leisure-related tourism (that portion of leisure-related tourism that is not visiting tourism/network-based tourism) are substantially, and therefore methodically, two essentially different elements. This is because of, among others, the following causes:

- For local recreation purposes, the challenge consists in making the best of the resources that one has (Clawson *et al.*, 1960). All larger concentrations of population (urban) in industrialized countries must have a minimum amount of free local space. Local recreational areas are to be an organic part of the city, irrespective of how outstanding they may or may not be.
- Leisure-related tourism (the part of tourism production that is subject to competition), on the other hand, is dependent on outstanding resources. The challenge is no longer to try to make the best out of what one has but to ascertain whether the resources are so outstanding that they are capable of forming a basis for production that can fare well in competition with alternative products located inside the country or abroad. The trip itself is an expression of the choice between alternatives.
- The local suburban free spaces have a relatively clear recreational function, whereas significant portions of leisure-related tourism have nothing at all to do with recreation in the traditional sense or with coverage of this type of human need.
- We may indeed encounter a situation where interesting local recreational resources in no way meet the needs of certain forms of leisure-related tourism and *vice versa*.

Needs Arise in People – Not in the Building and Construction Sector

Qualities, skills and artefacts are turned into resources only in relation to human needs and processes. Similarly, tourism resources only develop in relation to (leisure) tourists' current or potential needs and desires.

We cannot relate specific accommodation facilities to particular tourist resources, because accommodation facilities are unable to mirror the needs of the traveller. A hotel can just as well accommodate business travellers, or safari tourists, persons on a therapeutic stay after treatment in hospital, participants on a painting course or salmon fishermen. By the same token, a segmentation in relation to the length of stay, for example by focusing on week-long summer and winter tourism, or round-trip and short-break tourism, does little to understand an area's potential tourist resources. Primarily, accommodation facilities are locales for production. We might classify and describe the production facilities and the structure that the satisfied demand falls within, but this does not provide us with any information about whether:

1. the choice was optimal or first, second, third, fourth etc. ranked substitutions for something that could have satisfied the demand in a better way;
2. the demand provides us with no information as to whether consumers who declined to buy did so because what was offered was not suitable (the articulated but non-effectuated demand/ the latent demand); or
3. already known accommodation facilities and combinations of activities are an inadequate tool for product innovation (to track offers that can enhance the latent demand).

Product development and expansion of the market are designed to provide products that are able to:

1. satisfy already established demand in a more suitable manner; and
2. offer new products that can satisfy the needs that lie in the frustrated and latent demand.

If an analysis of resources for leisure-oriented tourism is to be conducted in a methodical manner, we must categorize the tourists into main types according to the motives and needs that caused them to travel. We can identify successful or less suitable tourism products based on the consumers' perspective, according to the effectiveness of the products to meet the travellers' wishes and needs. However, whether the market segment is large or small, affluent or poor, growing or declining, cannot be used as a criterion when evaluating whether the offer is good or poor for the given group in question.

Different Methodological Approaches and Experiences

Successful local development and appropriate choice of product for resource-based productions are dependent on our having adequate methods by which to find and analyse resources. If the methods are insufficient or primarily developed for a different purpose, there is a high probability that many of the resources present will remain undiscovered or wrongly identified.

Production of knowledge, such as research is, does not occur in a vacuum. Because of this, in the following section I will take account of the various methodical approaches and experiences in surveying resources for leisure-oriented travel and tourism.

The geographical method

The geographer Brian S. Duffield in *GeoJournal* (vol. 9, no. 1, 1984) sums up the 'state of the art' in the area of geography relating to the survey of resources for leisure-related tourism. In the section 'The evaluation of resources' he identifies that:

> It was inevitable that a major focus of the interest of geographers in tourism would be in the interaction of tourists and the resources they use. It was also to be expected that this interest would take the form of a functional and systematic appraisal. Such an orientation has long been evident within British geography and the early work of Comish (1934) on scenic amenity in Great Britain was an antecedent of a plethora of later studies based on a dual interest in countryside recreation and a concern for the preservation and conservation of scenic environments. In the more recent past, resource assessment for recreation have increasingly taken a functional form, recognising that resources are not absolute but dynamic and are defined by the nature and range of possible uses (Duffield and Owen, 1974b). As Zimmermann (1951, p. 7) has stated, 'the word "resource" is an abstraction reflecting human appraisal and relation to the function of operation'. Recreation resources were, therefore, seen as being defined culturally by the nature of the activities under consideration and from the late 1960s geographers struggled in an attempt to devise objective procedures to evaluate and measure

> the quality and value of landscape for recreational purposes...

We should especially take note of the following. In Duffield's summary he relies on the section 'Meaning and nature of resources' (Zimmermann, 1951, pp. 3–20), among other things, to point to the following very important traits of what we call resources:

- resources are entities that are not 'absolute but dynamic...';
- resources are an abstraction 'reflecting human appraisal and relation to the function...'; and
- resources must therefore be 'defined culturally'.

Duffield is on a very interesting track when he points to the cultural context. To cover more elementary needs (needs of necessity), resources will largely be the same across different cultures, but suitable resources to make provisions to cover needs of convenience and needs for luxury, in rising order, will be culturally dependent.[2,3]

To operationalize the cultural context in relation to tourism and resource analyses, we must continue in Duffield's footsteps and split our cultural reality into subcultures (i.e. what I elsewhere refer to as *segmentation*, and different *types of tourists*). To speak of some kind of 'Norwegian cultural uniformity' or 'Scandinavian cultural uniformity' etc. will not yield any results that can be applied. We need to limit the subcultures (the various types of tourists) so rigorously that we can ascertain *important common values, interests, indicators* etc.

Tourism resources and tourism products are given meaning and can be assessed qualitatively (assessed 'objectively') when analysed in relation to types of tourists sharing common traits in important areas. The next very important aspect in Duffield's summary is what he identifies as operating methods:

> Other techniques were developed and applied in a range of different geographical situations (for example, Blacksell and Gilg, 1975; Crofts, 1975) and techniques grew evermore sophisticated in terms of their utilisation of developments in statistical analysis and computing thus echoing contemporary trends in British geography as a whole
> (Coppock *et al.*, 1974; Coppock and Duffield, 1975).

> By the mid-1970s the various techniques that had been devised were already themselves the subject of critical review (Penning-Rowsell, 1973; Crofts and Cooke, 1974) and the objectivity of those based on the Linton approaches were being questioned
> (Gilg, 1975).

> Geographers have yet to overcome these difficulties and the last decade has seen little substantive work in this field, *an indication perhaps that solutions to this intractable problem lie outside the geographical discipline...*
> (Duffield, 1984, p. 32, my emphasis)

One possible answer to Duffield's question in this respect may be a sociological distribution of tourists into more homogeneous groups (types of tourists) so that common conceptions, interests, values and life views etc. clearly emerge. I shall pursue this line of enquiry later.

Duffield broaches another technical issue at the end of the final section by noting that:

> ... tourism studies have been predominantly of an empirical nature and conceptual and theoretical advance has inevitably tended to lag behind these empirical investigations...
> (Duffield, 1984, p. 32).

The academic situation internationally has principally remained unchanged since 1984. Within the geographical tradition, work still follows the same two main divisions: either landscapes and other potential resources for tourism purposes are surveyed under the premise that the major basis for leisure-oriented travel is recreation, or landscape is analysed on the basis of suitability for various activities. This, however, is a highly indirect way of ascertaining the needs, motives and driving forces that (the actual) demand comprises. This nexus can be visualized schematically (see Fig. 4.1).

In addition, mapping the suitability for activities and facilities provides merely an indirect connection to the needs of the tourist; it provides no preconditions for creativity. Instead, it establishes a process where what is known becomes the premise for identifying more of the same.

The landscape architecture tradition

Within the discipline of landscape architecture, there are primarily two main lines of enquiry that have been pursued. One of these is what

I shall call the *architect's perspective*, that is what we see as a division of labour between landscape architects and architects who build structures. In this tradition, the shaping of the rural landscape is the landscape architect's primary task, just as, correspondingly, the design of individual and more densely concentrated buildings will be the building architect's responsibility. The landscape architect has been challenged by major developmental projects such as power plant expansions, road and railway construction and major urban areas to be developed. In some cases, it has been a case of rectifying earlier landscape interventions. Like the work conducted by construction architects, the work of landscape architects can be approached from a resource-theoretical point of view. It can be analysed as management of resources, maintenance of resources and, possibly, resource development. The architect's perspective, however, cannot serve as an independent and all-encompassing method for resource mapping.

The other tradition in the field of landscape architecture has been to analyse various landscapes in relation to their *experiential value*. The model for this method is most frequently a method procedure developed by the US Forest Service in the USA (US Deptartment of Agriculture, 1974). The method has been developed as a basis for multi-purpose planning and it is used by most of the agency offices of the US Forest Service (US Forest Service, Alaska Region, Division of Recreation, 1979). One of the purposes for which this method has explicitly been applied is recreation. The method received much attention in Norwegian landscape-related academic environments; indeed, at the Department for Landscape Architecture at the Agricultural University of Norway (NHL), senior researchers have worked during the past two decades to develop and apply the method. This has been partly for use in assessing experiential quality in 'planned and potential hydroelectric power development' (Bruun, 1983; Geelmuyden and Berg, 1986). The method has been partly adapted for use in assessing the suitability of areas for tourism (Lykkja *et al.*, 1990; Svendsen, 1990; Lykkja, 1991). In an intuitive sense, this method may seem interesting in relation to tourism because it explicitly addresses recreation and experiential value. These are two key aspects of tourism.

How, then, is this method structured and what aspects will it enable us to encompass? The method is two-fold since, in the first stage of mapping, it requires the target area to be set in a landscape region if a smaller area is to be analysed. If it is a larger region or a county that is to be analysed then the areas should be divided into small units. The main point concerning '(a) division of the landscape into regions...' is that it 'provides opportunities to conduct landscape assessments within areas that are comparable...' (Svendsen, 1990, p. 2). The four criteria by which the natural landscape is divided into regions are:

- natural base
- land use
- building development and other cultural features
- total impression of openness – scale.

Coastal landscapes and mountain landscapes, for example, will thus ensure that they will each be included in the appropriate regional group. The unit that the landscape region then forms will, in turn, form the basis for ranking the areas in the region according to a three-item scale from A to C.

> Class B represents the areas that have a typical landscape based on the characteristic landscape

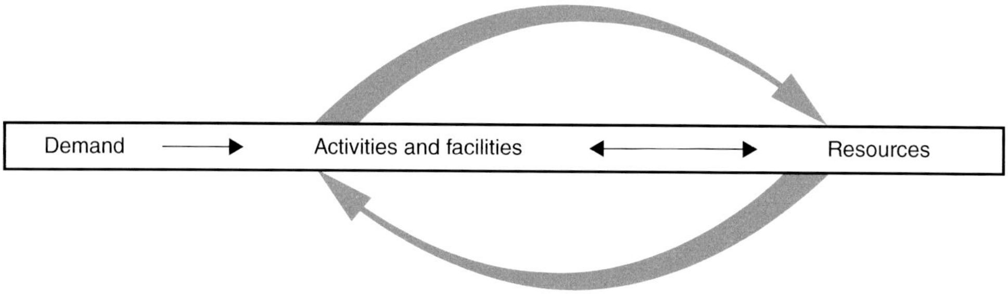

Fig. 4.1. Relationship between demand, activities and facilities, and resources.

> traits in the individual landscape region. Class A encompasses landscape that is particularly varied and distinctive in terms of the landscape that is typical of the landscape region. Class C accounts for areas that fall outside Classes A and B, so that Class C comprises areas with landscapes that have little variation.
>
> (Svendsen, 1990, p. 3)

The next stage in this method is to analyse 'the experiential value of the landscape'. The criteria to which the method links experiential value include:

- aesthetic value (keywords: landscape's linear patterns, landmarks and cultural heritage/ relics of significant importance);
- symbolic value (keywords: important historical events or persons, activity etc.); and
- identity value (recognition value) (keywords: cultural heritage that embodies a social, historical and ethnic affiliation for people).

Next comes an item of great importance to the method (Svendsen, 1990, pp. 11–12):

> In conjunction with a method that pertains to assessing the experiential value of the landscape, it is particularly appropriate to emphasize the aesthetic aspect of the phrase experiential value. ... Research in recent years stresses three basic criteria as important in the assessment of the landscape's experiential value. These three criteria are (A2)

- the whole
- variation
- intensity

The value parameters in the penultimate point above, although attributing general value to the aesthetics of landscape, seem to me to be quite problematic, if not incongruous, based on what we know about various types of tourists.

A few examples would be useful here. The Gothenburg area has become an important alpine skiing market for Hemsedal, because it has been linked with site-specific aspects: *topography* (sufficiently challenging trail routes) and *temperature* (a qualitatively good product is ensured during one's stay). Malleable resources include: a level of *knowledge-related competence* so that the trail routes are well prepared (the raw materials being topography and precipitation in the form of snow) and have been processed so that the *main product* maintains a high quality. Of course, an additional quality can be the surroundings having good visual quality, but they do not represent 'the more fundamental resources', to borrow Travis's words once again (Travis, 1982, p. 257). The tourist businesses of Espedalen in Sør-Fron Municipality in Oppland county have developed their area into one of Norway's foremost tourist destinations for skiing, which is the training centre for the elite of Norwegian cross-country and biathlon skiiers. This has been due to the area having a few crucial input factors of 'outstanding quality', which were present before these developments. In terms of natural assets, one can place at the top of the list *suitable topography*, a long and predictable winter (with precipitation in the form of *snow*) and vegetation (*flora component*) that provides shelter from wind and storms. In other words, this provides confidence and stability in the conditions for skiing. The second crucial production factor consists of *competences* and *networks*. 'Networks' refers to the user networks and also the sports, health and nutritionist environments. In terms of *landscape aesthetics* and *landscape variation*, however, Espedalen is surpassed by a large number of other areas in Norway.

There is only one prerequisite that motivates a passionate salmon angler to travel halfway round the globe to practise what the British in the previous century called such 'a fine art'. Good climate conditions are secondary; beautiful natural surroundings are subordinate; and other natural aspects are inferior in importance. The sole factor that counts is *a single occurrence within the fauna component*, namely, the probability of hooking a sizeable salmon.

Sport fishing from ocean-going vessels (one of the larger organizations for organized fishing is the Federation of Sea Angling Clubs) is primarily oriented towards catching the true giants such as the Greenland shark, porbeagle, spiny dogfish, halibut and various larger fish, or what one technically terms 'specimen fishing'. The latter has also been called wreck fishing, because it largely involves fishing around shipwrecks and similar ocean-floor situations that are attractive spots in which special species of fish like to congregate. Deep-sea fishermen also travel the globe to the same extent as passionate salmon fishermen and this is due *exclusively to the fauna resource component*. To find and identify presumably interesting areas, a knowledge of *underwater topography*, *ocean currents* and *special ocean-floor conditions* is important:

> [O]ne may wonder about Norway's place in the development of ocean fishing tourism. While one can find a large number of sport fishing centres throughout the world that have based their existence on merely a few species of fish, these places normally require one to reserve a fishing spot far in advance. Norway has little or nothing at all to offer… [Still,] We have fish that are distinctive in terms of amount and species when compared with any other country in the world.
>
> (Søyland, 1985, p. 79)[4]

Perhaps tourism operators are unaware of the available resources because the methods that have been applied for analysis are inadequate for discovering these types of tourism resources.

The fact that 19% of all tourists visiting Israel go there 'to visit places that have religious association' makes basic natural conditions, synthetically created attractions and operator expertise at the production site largely irrelevant. The main driver for this type of tourism is not primarily nature and its landscape. The driver is non-material and derives from the fact that the *area is one of the crucibles of western European culture*. Leisure travellers in the form of 'to visit friends and relatives' account for 33% of people visiting Israel, inferring *network-related tourism*, not resource-related (ibid.). Furthermore, 22% of tourists visiting Israel list sun and sea as the main reason; they are the *sun worshippers* and, using Aubert's categorization, it appears that the conditions for this are linked to *a few basic natural components*.

Using the relatively new city of New York as an example, it has become one of the great meeting places for the modern 'jet set' (a significant group in our age, the *work-free class*, if we borrow one of the terms deriving from Thorstein Bonde Veblen (see his main work from 1899)), and it is not due to any natural resources or conditions, but instead to the fact that the urban centre is one of the *great producers of western culture* in the modern era.

Another example could be Nepal, which in recent decades has had a volume of tourism that is on the verge of causing ecological imbalance, largely due to the presence of the world's tallest mountain (Everest); in other words, a *geological and topographical resource*. An added resource, for many visitors, is the people who live here, and their religious relationship to the mountain ranges, which are also important reasons for tourists to visit the area, i.e. aspects that are *culturally related* and of a *non-material nature*.

> One does not travel to Poland to experience beautiful, colourful cities and to sit at fashionable pavement cafés. Poland has neither of these […] Poland's primary attraction is the Polish people's national character, systematically and perpetually tormented throughout history. An emotional, reliable and Catholic nation on the European map, squeezed between two dispassionate superpowers – the orthodox Russia and the meticulously well-ordered Germany.
>
> (Bech, in *Politiken*, 1983)

'The national character of the people is an attraction in itself', according to the author. If this is true, Poland is attractive to tourists because of its historical past and its geographical position in history. The attraction, according to the author, is socially related and non-material in nature and different to the examples above.

The 'state of the art' pertaining to tourism in Norway

The most recent professional overview of this topic dates from 1990, when the report entitled *Norway Tourism* was published. The Travel Business and Areas was a method for mapping and presenting travel and tourism's areas of interest (Lykkja *et al.*, 1990). The report was prepared by the Norwegian Landscape and Forest Mapping Institute (NIJOS) and commissioned by the Ministry of Trade, Tourism Department. This organization is a fact-finding and research institute under the Ministry of Agriculture. The Norwegian University of Life Sciences (NMBU), Department of Landscape Architecture, has taken an important lead in the use of this method which was adapted in this department (described earlier in the chapter as a method used by the US National Forest Service for its multi-purpose planning). The main co-ordinators were the geographers in the tourism section of the university; however, other disciplines were also involved including *landscape architecture* and *natural sciences*. The method will be tested against four key indicators:

- academic integrity;
- knowledge-generating potential of the work;
- basic theories of tourism and society; and
- the starting point in production theory.

Academic integrity

The Norwegian Forest and Landscape Institute has undertaken the mission as a typical commissioned, fact-finding institute that has gone to great lengths to accommodate the commissioning authority (tourism section of the Ministry of Trade). Therefore, the commissioned institute has not mustered the courage to raise professional objections but, on the contrary, has incorporated NIJOS into its working methods, policy myths and false notions. From the very outset, for example, we meet the following policy myth, which, from a professional perspective, is entirely implausible. 'Tourism is selected out as a growth industry' (p. 11). In terms of policy, one can select *areas of commitment* but not *growth industries*! The latter lie outside what is politically possible in a society comprising largely private enterprises, an open economy and free competition to a significant degree. Why were earlier Norwegian standard manufacturing industries not cited as growth industries – the garment industry, the footwear industry, the canning industry, the electrical appliance industry or the standard shipbuilding industry? All of these are examples of market sectors and products in growing global demand and all of them were rurally localized industries.

Knowledge-generating potential of the work

The main objective of the project was to develop a suitability chart as an aid in tourism planning. The commissioned institute reveals in its report that it is not familiar with earlier Norwegian efforts to address the challenge of producing a suitability and resource chart for tourism planning. At the end of the 1960s and the beginning of the 1970s, this was a professional challenge in Norway that several of our most competent individuals and expert groups became engaged in trying to address. Both the individual theme maps and the mapping system itself were developed during this period and went considerably further than what the Norwegian Forest and Landscape Institute came up with after about three years' work (cf. Langdalen and Stav, 1974; Meistad, 1974). If one places it in a professional context of work done in Norway, the project adds no new knowledge; it succeeds only in bringing new players and institutions/departments into the field.

Basic theories of tourism and society

The project is out of touch with the 'state of the art' in terms of tourism and society. A suitable place to begin to demonstrate this is page 15 in the report, where it is claimed, without reservation, that 'Tourists are normal everyday people who, in their leisure time or holidays, want to do something different from what they usually do...'. In this respect, NIJOS fortifies two myths that policymakers have deemed very important to bear in mind. First, that holidays and holiday trip-taking are universally allocated benefits ('Tourists are normal everyday people...'). This is absolutely wrong, because holidays and trip-taking are clearly associated with people's working situation and income (Statistics Norway, 1983, pp. 280–286 and Statistics Norway, 1989, pp. 176–178). Secondly, we know that there is a close connection between one's work situation and holiday behaviour. As a group, non-professional employees are the only people who can be said to see leisure as the grand opportunity to compensate for a difficult work situation, poor living and neighbourhood conditions, or a bad start in life because of minimal schooling or absence of other training. In the Danish working environment found in *Youth, Education and Leisure* (Fredriksen and Møller, 1977, pp. 25, 26, 121), the following connections are emphasized:

> The aforementioned fragmentation of everyday life is expressed in discussions about 'leisure time', but often in a manner that may appear to be ideologically distorted. One may encounter the notion that it makes little difference that one has a lowly education and an uninteresting job, because 'on holiday one can do exactly what one

> pleases'. A notion of this kind is based on the supposition that leisure time can compensate for almost anything at all...
>
> In the following section, we shall present the empirical data that shows that the above-mentioned notion of leisure time lacks a basis in material and social reality...
>
> Leisure time can be meaningful for most people, if their work is meaningful...
>
> Sociologist professor John R. Kelly (1987, pp. 177–179) points out the same idea in the following:
>
> Social stratification has been a fundamental theme of sociology from its founding. Modern post-industrial societies are characterized by divisions...'.
>
> Max Weber (1958)[5] offered an enduring distinction that will be followed in this discussion. He defined social class as 'life chances' – placement in the socioeconomic system determining access to resources and opportunities...'
>
> The reward system operates to attract individuals into economic positions and behaviours as the labour and capital markets require. Leisure is defined as reward for functional behaviours that are supported out of economic surplus. Leisure, then, is earned rather than being a human condition or right...'.

For additional examples, see Parker (1965, pp. 65–75).

Returning to the report, page 15: the tourist will '...usually get in contact with the permanent local residents and with other guests. They will experience something authentic, feel secure. And they will usually feel welcome among the tourist hosts and the permanent residents...'.

To recap, the main groups from the segmentation work of Erich Cohen divided tourists (international tourism) into the following main groups:

- the organized mass tourist
- the individual mass tourist
- the discoverer
- the drifter.

In this respect, note the following: on a global basis, the majority of international tourists fall into the first two groups. In Cohen's categorization, it is not until we get to the third main group (the discoverers) that we meet a tourist with an apparent desire to come into 'contact with the local permanent residents' and to experience 'something authentic'.

On page 18, the report describes tourist enterprises (hotels and dining places) as tourism's infrastructure (emphasized in quotation marks). Here again, NIJOS is at odds with internationally acknowledged terminology. Tourism researcher Anthony S. Travis denotes these types of facilities as tourism's suprastructure (Travis, 1982). The latter conforms exactly with the terminology used, for example, by Forschungsinstittut fûr Freizeit und Tourismus, University of Bern, Switzerland (Kaspar, 1975, p. 52). Both words in tourism are used as technical expressions; tourism infrastructure and tourism suprastructure. Overnight accommodation and dining establishments are tourism suprastructures and dependent on the production apparatus having an environmental and experiential value associated with them (they are something more than traditional production facilities).

The starting point in production theory

On page 21, the unreserved claim is made that 'in a composite tourism plan, it is necessary to conduct a market survey to ascertain which products are in demand'. This may be so, but it covers only partially what must be ascertained. What is generally accepted in production theory is that there are the following three main forms of demand:

- effectuated demand (articulated and satisfied demand);
- frustrated demand (articulated but not satisfied demand); and
- latent demand (non-articulated demand).

Market surveys can provide important information pertaining to *product improvements* and whether the market will accept *product copies* but will provide no basis for developing latent demand. Product development can and should occur based on information from the market, but product ranges occur on the basis of the product developer's understanding of demand from the inside (the needs, motives and driving forces that result in demand). Understanding producers from within their knowledge base will help develop new and better product responses. Norwegian tourism is currently in the situation that necessitates product range development. The conditions for this are built up through competence within professional disciplines such as recreational sociology, recreational geography, social and cultural history, consumer theory, production theory, knowledge of raw materials etc.

When placed in an international academic context, the NJIOS report belongs to achievements of the 1960s. In a Norwegian academic context, it is clearly inferior to several of the earlier mentioned works in Norway during the 1970s. There is a set of justifications behind this development (as we know, development is not a linear movement forward; in terms of both culture (wholes) and knowledge and skills (in individual areas), deterioration can occur). The most important function in development, however, is quite apparent.

First, the social science disciplines provide approaches to understanding needs and motivations of the tourist. These differ from a landscape architectural approach that uses value parameters such as *aesthetic value*, *symbolic value* and *identify value*, which are parameters that are place-specific. There is a very high degree of local area parameters. Hence, these parameters are meaningful for groups of local and regional users (or to the experts who unconsciously or consciously defined the content of those parameters) and randomly, for subcultures and distant users who have another cultural, historical and social background.

Secondly, tourism has become a field of academic interest and institutionalization during the past two to three decades. Since resources are something that exist in relation to human needs and motivational forces, working with resources for leisure tourism should depart from the academic discipline of tourism in combination with the academic disciplines mentioned above. In this respect, tourism is like all other academic fields: there are no 'short cuts' like those the Norwegian Forest and Landscape Institute has attempted to use.

Thirdly, building competence and attaining a position allowing interaction with researchers entails targeted and hard work. Professor Rolv Mikkel Blakar, epitomizes this by saying, 'To qualify for the research front, one needs 15–20 years of continual education, effort and qualifying work. Only then ... can one hope to contribute seriously by providing new knowledge...' (Blakar, 1989, pp. 1, 3). Those who wish to reduce the criterion of knowledge to a fact-finding work such as the one submitted by NJIOS will gain no acclaim from Professor Blakar, since he emphasizes that 'the applied research must fulfil the same requirements as those stipulated for basic research...' (1989, pp. 1, 3). It is suggested, given the above, that a research process of three years – this being the first time NJIOS was commissioned to investigate tourism – cannot result in anything other than a report stating that 'work on a theme has been done'.

Conclusions from the review of methodology

Sometimes research assessing tourism resources, using geographical methods, is only partly successful, for which there are a number of reasons. First, as previously noted, geographical research methods are primarily designed to map and assess *local recreational resources*, which are contextually different from tourism which has a leisure-recreational motive. Secondly, geography will be insufficient as an independent approach to explain *subcultures*, or *tourist types*, which define motives, needs and conceptions linked to identifying and locating the resource components of tourism. Furthermore, if it is true that resources exist in relation to human needs and if it is also true that reality consists of various cultures in a global context and subcultures in local areas, it is not so surprising that Duffield ends up concluding that '...perhaps ... solutions to this intractable problem lie outside the geographical discipline...' (Duffield, 1984, p. 28).

The method offered by the landscape architect for mapping tourism resources is of limited value for tourism, the reasons being similar to those already identified with the geographical method. First, the method I have described here was developed for assessments in various fields, not just tourism. One of these purposes has been recreation, not local recreation but rather *distant recreation for the region's own residents.* The concept of recreation used here is also largely synonymous with *outdoor recreation*. Furthermore, the value parameters – aesthetic, symbolic and identity values – are related to a region's shared values. This can be meaningful for a majority in a region, but human needs and values that are present in local subcultures or in society outside the region are excluded. The other significant reason why the method offered by the landscape architect does not have the potential for a complete resource analysis is the dominant role of aesthetics.

A common element in both the geographical method and the aforementioned landscape architecture method is the fact that both these methods are based upon natural conditions and visual factors. This means that very important parts of the socially determined resource, that is the non-material resource, will not be adequately represented. Secondly, both have a significant ethnocentric bias, particularly in the landscape architecture method.

This content of experiences fits only a few of the tourist types I have described earlier from the 'tourist typologies', which I will discuss in the next section based upon interpreting and expanding Aubert's typology. In addition, the method offered by the landscape architect is too imprecise for evaluating resources in relation to the many specific examples of set-ups of tourism and resource components presented earlier, in the section 'Tourist typologies'.

In addition, the method used by landscape architecture is imprecise when it comes to evaluating resources. It also lacks an ability to contribute to the explanatory value which would be useful in the many examples of tourism and its supporting resource elements. I have presented these in the section on 'Tourist typologies'.

My working hypothesis to identify tourism resources is as follows: the geographic method must be combined with sociological understanding. We must travel via sociology to gain access to a tourist typology that distinguishes tourists from one another according to different needs, desires, life views and motivations. By using another discipline (sociology) and by conceiving of leisure-oriented tourism as something other than recreation, we have established an interesting point of departure to develop an alternative model of *resource identification and resource assessment.*

Interpreting and Expanding Aubert's Typology

Aubert's (1969) work, which I presented earlier, was published during the latter half of the 1960s. In the course of the 70s and 80s, changes in values and knowledge within society generated new demand groups. I have therefore found it necessary to expand Aubert's typology to include two new main groups.

I have chosen to call one of these the *activity tourist.* For this category, the connection can be traced back to the 'jogging trend' that occurred in the USA during the early 1960s and the tourist seeking physical activities for exercise with like-minded persons.

I call the other group the *utility tourist.* This group, like several of Aubert's six main groups, comprises two different subgroups. The first subgroup to appear (explained by having a private car) are those who want a reasonable change of environment and at the same time the chance to utilize their holiday for some productive benefit by spending a portion of it to gather stocks of food and fuel to lighten the burden on family finances. During the 1980s, a new subgroup surfaced. It consists of people who wish to supplement their household pantry with fruit, berries, fish and wild game, all of which are free of chemicals, hormones and other artificial ingredients. In order to ensure that the interpretation of Aubert's tourist typology is adequate for both this chapter (on resource analyses) and for the final chapter (on adaptation of tourism in local society), I shall examine each tourist group in the areas of:

a) what the group is seeking (summary from Aubert);
b) activities and facilities that satisfy the group (added by me);
c) energy consumption, pollution and costs (added by me); and
d) adaptation to local nature and local culture (added by me).

Elaborated and interpreted in my system, Aubert's work provides key information for understanding how tourism works in local society.

Vilhelm Aubert conducted his analysis, as I have already indicated, at the end of the 1960s. The main focus for using tourist typologies in our context is to understand the characteristic tourist roles and their needs, motives and social connections. Tourist types are thus time-determined to a certain extent, because what is useful at a particular time is determined by values, knowledge, skills, finances, technical possibilities

1. The anti-tourist	
a) Seeks[6]	This is the leisure traveller who does not wish to be perceived as a tourist, but rather wants to be a social and cultural consumer in the local society where he stays temporarily. The category 'anti-tourist' is principally composed of two groups that encounter local society with entirely different circumstances. The majority in this group are people who have moved away from their home location and who seek to return to their roots during their holidays. Socially, culturally and interactively, they easily fit into and become part of the local community during their stay. This segment of the group therefore falls outside the sociological definition of a tourist such as delimited in Cohen's work from 1974. The other group (in terms of visiting a locality belonging to the resource-based tourism group – see the same reference as above) are visitors who wish to adapt to the local society both socially and materially during their visit. Because the stay is of limited duration, this adaptation may be only partial, but in our context the need to integrate is an essential part of this type of tourist. The visitor does not deliberately demonstrate differing values, morals or attitudes. The visitor strives for cultural approval.
b) Activities and facilities that satisfy the group	For those who spend their holiday returning to their original cultural roots, it is usual to stay with close relatives or next of kin, or in their own holiday residence. For many these are ordinary cabins, but quite often they are older houses that they have taken over or purchased in their holiday community. The others in this group, those who are in a real sense (not only formally and statistically) strangers, are nevertheless 'determined to take part in significant aspects of life among the permanent dwellers' and therefore adapt materially (Aubert, 1969, p. 139). They rent quarters in boarding houses, private homes, houses and cabins.
c) Energy consumption, pollution and costs	This is a minor resource-intensive form of tourism, partly because the tourists stay in one place and partly because this form requires little extra effort/resources on the part of the receiving area. Existing buildings are used to a large extent and the 'anti-tourist' adapts to the traditional local food, beverages and entertainment.
d) Adaptation to local nature and local culture	The anti-tourist is a sensitive consumer in terms of nature and culture. This is because the resources he needs have a low impact. This type of tourist is responsible for important transactions between society at large and the local community.
2. The outdoor tourist	
a) Seeks	This is the tourist who turns his holiday into a completely different setting from that of his home location. First, the outdoor tourist escapes daily social obligations and everyday conformity. Second, the outdoor tourist seeks physical and mental challenges. The holiday trip requires meticulous planning and the participants are thoroughly resourceful, financially, socially and cognitively. The participants come from the political, academic and business elite in the highly industrialized parts of the world.
b) Activities and facilities that satisfy the group	What is important for this type of tourist is to be closely integrated with their natural surroundings and create new challenges. The outdoor tourist wants to be able to manage on his/her own during the trip. Accommodation is primitive and rudimentary – tents, sailboats or unattended campsites. Comfort during the holiday trip will contrast with the homey atmosphere of where they come from.
c) Energy consumption, pollution and costs	The holiday trip (expedition) is frequently extremely energy-intensive and costly, if one includes the entire project including production of equipment and the long trip (western Europeans and Americans travelling to Nepal/Asia, trips arranged by the Norwegian Trekking Association (DNT) to Kilimanjaro in Africa, trips to the South Pole area currently being planned by the Swedish Tourist Association, skiing expeditions to Nustagh Ata in Asia (China), or the classic sea regattas such as Whitbread Round the World Race).

Continued

d) Adaptation to local nature and local culture	The outdoor tourist normally seeks natural areas without major local human habitation and adaptations. There local culture and local nature are considered as undisturbed, but today this is only true in rare cases. The sustainability of alpine and polar areas is fragile so that what nature is able to tolerate can be quickly exceeded. Tourism involving expeditions to Nepal are a striking example of how local nature and local culture's sustainability has been encroached upon for many years (Bjønness, 1980, pp. 119–138). In other cultural contexts, the conquering and popularization of natural components may be seen by the locals as a desecration of holy symbols and spheres: '... in the Hindu as well as the Buddhist religion the major snow-covered peaks in the Himalayan Range are sacred mountains...'; 'Climbing the Khumbu Yul Lha is seen as a threat to the Gods and many of the Buddhists would be upset if westerners set foot on the sacred mountain' (Bjønness, 1980, p. 133).
3. The emigrant tourist	
a) Seeks	This is the tourist who '...spends a significant portion of his time away from his native country because he is against certain aspects of its societal structure' (Aubert, 1969, p. 118). The motive here is more to get away from aspects that one dislikes intensely; for example, a too-strictly regulated society, a society that is not free, a too-transparent society etc. In a social context, it is often the resourceful 'freeloaders' who are not a part of the binding social interaction, neither at home nor in the niche environments abroad. 'Since their principal motive is to get away, they seek only a comfortable and cheap place to practise their tourism, without worrying to any extent about the qualities of the receiving society aside from prices, taxes and the weather' (Aubert, 1969, p. 118).
b) Activities and facilities that satisfy the group	Aubert's emigrant tourist category obviously comprises two essentially different groups. One of these groups consists of the conservative businessman and the reactionary citizen. These exist based on capital earned in their home country. The other subgroup encompasses 'the young and somewhat deviant emigrant tourist. The group is not engaged in important transactions with the local population ... but lives inexpensively and often takes random and short-term jobs. Characteristic representatives of this group are often artists, semi-artists, alcoholics or sexual deviants, or merely a young person who drifts about' (Aubert, 1969, p.132). Cf. Cohen's term 'the drifter' (Cohen, 1972, p. 168). This group most often lives simply and casually; they lack capital and they make their living by selling labour.
c) Energy consumption, pollution and costs	The emigrant tourist remains stationary for longer periods and typically lives simply. Staying in-place coupled with a simple way of life yields low energy consumption.
d) Adaptation to local nature and local culture	For these reasons, the emigrant tourist is a sensitive tourist in relation to local nature and culture, primarily because the emigrant tourist, to a greater extent than most other tourist types, brings with him an everyday role and is not a 'big spender'.
4. The socializer/sightseeing tourist	
a) Seeks	'For this category of tourist, the journey per se is not so important' (Aubert, 1969, p. 136). What is essential for this voluminous, large group is the social relationships with fellow tourists or the social opportunities that pictures and experiences offer to the tourist's social status after the trip. Nature and culture are secondary but, nevertheless, important backdrops.
b) Activities and facilities that satisfy the group	The main activity for this tourist type consists of being moved in space in a socially beneficial manner. Cruise ships, yachts and touring coaches are the classic means of transport and museums, grand sceneries, metropolises, picturesque villages, cultural crucibles etc. are the key 'collector's items'.

Continued

c) Energy consumption, pollution and costs	Daily consumption for the socializing tourist has traditionally been high, and generally in this group one finds lots of: • abundant free time (one cruise is succeeded by another); • abundant consumption (traditionally so costly a form of holiday that it is the exclusive domain of the wealthy or otherwise privileged citizens of society; cf. those eastern Europeans who have comprised the bulk of this group); and • abundant leisure skills; this means insight into social customs, norms and activities. In this group, material consumption is high both at home and in the tourist situation.
d) Adaptation to local nature and local culture	Local nature and local culture are consumed by this tourist type in a passive manner. Picture-postcards, the eye of the camera and exclusive souvenirs are the link between the traveller and the place visited. Learning is minimal; the exotic, simple and stereotypical is what is internalized.
5. The expert/connoisseur	
a) Seeks	This is a category of tourist who has special interests cultivated during their holiday. 'They are interested in an aspect of the receiving society and its population, such as folklore, music, handicrafts and applied arts, or very simply their everyday customs and the social and economic conditions in the host country' (Aubert, 1969, p. 134). The expert often has a professional/academic background and spends his holiday cultivating interests that his working situation does not allow. The holiday becomes an 'extension' to accommodate important personal interests in the tourist's everyday life. (For the term, see Parker, 1965.)
b) Activities and facilities that satisfy the group	What members of this group share is a quest for the outstanding and genuine. These can be found and experienced in both nature and culture. The experience itself may be physical and freely accessible (botanical, geological or other natural phenomena), or it may be attendance at costly cultural and social events. Typical attractions for this group are visits to the Salzburg Festival, the Wagner Opera Festival Theatre in Bayreuth, the Cannes Film Festival, theatre weekends in London and New York, jazz fests etc. The 'connoisseur-tourist treats the world as a cafeteria of curiosities where he can gorge himself and his home country on selected dishes' (Aubert, 1969, p. 135). The participants in this group are mainly recruited from the same social strata as outdoor tourists and they are often the same persons moving between the two tourist roles.
c) Energy consumption, pollution and costs	The connoisseur is often out travelling and often goes far for brief visits. The many long-distance journeys use a large amount of energy and daily consumption may frequently be high. As a collector, he is often cynical about acquiring plundered and illegal antiques, animals and plants.
d) Adaptation to local nature and local culture	The expert is a collector of specialities, very frequently without time for, or interested in, cultural and social encounters. He acts often on his own premises, and practises his own values, norms and material standards.
6. The sun worshipper	
a) Seeks	'This is a tourist who primarily chooses an area for climate benefits (dry and warm climate for summer activities, and stable, dry and sunny winter climate for sports and recreation) ... combined with the desire to live the good life' (Aubert, 1969, p. 117). The sun worshipper seeks the company of other tourists. The charter tourist heading south is a classic example. In Norway, this is the tourism found at major winter sports facilities and in summer along the coastal strip in southern Norway. The role they play in everyday life is, for most of them, quite different from their role as tourists.
b) Activities and facilities that satisfy the group	This is the tourist type who provides the basis for the large tourist ghettos and amusement parks (the non-site-specific, synthetically created ones). For this category of tourists, entire areas are tailor-made including hotels, bungalows, restaurants, dining establishments, amusement park facilities, guided tours, transport etc. Charter tourism to the Mediterranean areas during summer and winter tourism to the Alps and the large winter sports places in the Nordic countries serve as typical examples.

Continued

c) Energy consumption, pollution and costs	The sun tourist is a mass phenomenon (cf. 'The organized mass tourist' in Cohen, 1972, p. 167). This is the holiday type of the masses *par excellence*. It requires little energy in the areas of accommodation and transport (for specific figures on energy consumption pertaining to the different means of transport see Kruschke, 1974, and Müller, 1992). Daily energy consumption has a low to medium output.
d) Adaptation to local nature and local culture	The sun worshipper has little or no adaptation to local culture and local nature, partly because he is unable to do this and partly because the holiday site is used as a cultural and socially free port. This tourist disassociates himself from his everyday role while at the same time playing out a role in his holiday that is socially and culturally disconnected and without obligation to the receiving area.

7. The activity tourist

a) Seeks	This tourist seeks specific leisure activities and the security that is found in participating in something that many others do. The activity tourist is a child of the jogging trend that began in the USA in the early 1960s. The outdoor tourist and the activity tourist are on opposite sides of the exercise scale, in approximately the same manner as the expert and the sun worshipper are on opposite ends of the 'knowledge-about-the-receiving-area' scale. Consider the following introduction to the classic sun worshipper/charter tourist (Cohen, 1972, p. 169): 'Where were you last summer?' 'In Majorca.' 'Where is that?' 'I don't know, I flew there.' (Conversation between two girls, published in a German periodical)
b) Activities and facilities that satisfy the group	This type of tourist has simple needs. The receiving society is of no interest. What he seeks is a schedule of activities, equipment and a programme that offers him opportunities to exercise suitable for his age, skills level and physical condition.
c) Energy consumption, pollution and costs	The activity tourist likes to stay in place. This form of holiday requires little equipment so that daily use is low to medium. Energy needs are also moderate.
d) Adaptation to local nature and local culture	The activity tourist has little interest in the receiving area, and he does not want to experience the unique and fragile environments either in nature nor culture.

8. The utility tourist

a) Seeks	This tourist is a composite of two subgroups. The one group comprises mainly utility tourists based upon an inexpensive holiday and reduced living costs by gathering and storing food; for example, collecting fish, berries, mushrooms and firewood etc. The other, and rapidly developing, subgroup are those who become utility tourists because they want to provide their families with pure food that is free of chemicals, artificial fertilizers, crop sprays and medicinal additives etc.
b) Activities and facilities that satisfy the group	These, in a way, are tourists who advocate a 'back to nature' philosophy. The type of accommodation and lifestyle is therefore best described as simple living. However, this group must have access to a number of technical aids in the form of canning facilities, tools and other necessary instruments.
c) Energy consumption, pollution and costs	This is also a type of tourist who is mainly inactive. Daily consumption is low. Functionally, much of the daily routine is a derivative of leisure.
d) Adaptation to local nature and local culture	The utility tourist has a positive attitude towards nature and local culture. He is inquisitive and somewhat critical of consumer society and he often exhibits positive humanistic and ecological views on life.

etc. The two new tourist types previously described have emerged over the course of the last 25 years.

The Sociologically Based Method of Resource Mapping

The following discussion will present the sociologically based method of resource mapping which will provide direct links between needs, resources and products (Fig. 4.2) (Table 4.1).

In terms of methodology, this represents a fundamental difference from the scheme presented earlier. The middle link – activities and facilities – along with the problems of interpretation they represent, has been removed. Instead, we now have a method that compels us to focus on the needs side (various tourist types) to a degree that enables us to understand the tourists' demand from the inside. Rather than describing different expressions of demand, we have worked our way to insight into the needs, processes, driving forces and motives that trigger demand. In principle, the method enables us to determine what may be a suitable product for the tourist.

To illustrate and demonstrate the sociologically based method for resource mapping, I shall conduct a rough macro analysis of the resource potential in Norway of resource-based tourism.

I will use Aubert's tourist typology in the model. In the example that follows, I have also:

- added descriptions of motivation and needs to the examples of buildings and constructions (elements of tourism's suprastructure) that can be used for these tourist groups;
- added comments on the Norwegian situation regarding resources and knowledge requirements; and
- proposed assumptions regarding the market and its development.

Summary

In order to understand the problems and challenges of different available resource-mapping methods, I choose to depart from Vilhelm Aubert who answered the question: 'What is a method?' in the following manner:

> A method is a way of proceeding, a means to solve problems and arrive at new knowledge. Any means that serves this purpose belongs in the arsenal of methods.'
>
> (Hellevik, 1991, p. 14)

First, the sociological method represents a clearly alternative process to the geographical and landscape architecture methods. Through the sociological method one obtains important principles:

- there develops a direct link between needs and resources;
- included is the entire leisure spectrum from the purely recreational to the extremely performance-based, and from the socially and culturally near and mutual (anti-tourist) to the interpersonal distancing and class-differentiated (forms of genuinely symbolic consumption), if one chooses this as the main approach; and

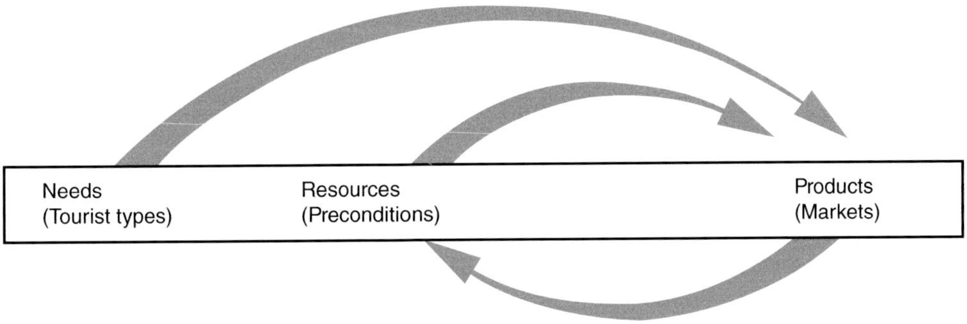

Fig. 4.2. Relationship between needs, resources and products.

Table 4.1. System for the sociological method.

Tourist type (needs)	A. Important resources B. Infra- and suprastructure	Product solution	A. The Norwegian resources situation B. The market
The OUTDOOR TOURIST is one who wishes to: – get away from one's own society – not to be hosted by any host society (wants in a sociological sense 'to manage their own' situation) – experience nature close-up – learn to know oneself better – avoid social conformity	A. For this group, basic nature is important. Continuous pristine areas for certain more demanding forms of outdoor tourism (cf. the term Qualified wilderness in NOU, 1974, pp. 28–29 and SOU, 1971, pp. 200–201 and pp. 381–389). Other forms of outdoor tourism require clean seas (diving), suitable mountain terrain having good hunting and fishing opportunities B. Varied desires and minimum demands on social infrastructure and tourism-related supra structure. The classic outdoor tourist has few or no demands in this area, whereas those who practise diving and sailing most often have certain minimum demands in both.	– sea sailing – mountain climbing – glacier trekking – river and rapids paddling – scuba diving – expeditions and safaris to distant and hard-to-access areas – trips on foot in wilderness areas	A. The Norwegian (and Nordic) resource potential is significant in an international context but is insufficiently analysed. Knowledge-intensity to develop marketable products. Demands first-hand knowledge about nature experiences and interplay between people and nature. This is competence we have failed to adequately develop in Norway and the Nordic countries for product development for the tourist market. B. The market is rapidly growing internationally; it is affluent and willing to pay.
THE ANTI-TOURIST is looking to: – not be identified as a tourist – not have 'tourist specific' demands from the receiving society neither material nor in terms of service – be integrated in the normal social activities in the society he visits, i.e. he wishes to be a part of the social network and local social life	A. The dominant share in this group has no special demands on resources because it is dominated by people who moved away in the first and second generation. Visiting family, friends and relatives (in terms of localization from main group network-based tourism). For the others, cultural and social aspects in the host area are important. The host location has minimal commercial benefit from the tourist. B. No demands on societal infrastructure and tourism suprastructure because the host society itself is the 'non-tourist-adapted form'.	– cabin in home community – renovated residences – mountain farmhouse, fishermen's cabin etc. – holiday stay on farms – rent rooms in private homes – boarding houses	A. The majority of vistors (network-based tourism) cannot be linked to tourism-generating resources, Norway's high cost of living and culture resources have only minimum appeal. Less industrialized areas of Europe with lower prices, warmer climate (and often Catholic background) have more to offer this group. B. The market (the resource-based part) in a global context has identifiable groups: intellectuals, pensioners and the well-off). The market is in growth in a global sense.

THE EMIGRANT TOURIST is one who is looking to: – find a society that is comfortable and cheap without a need of worrying about the host's society's conditions beyond prices, taxes and the weather (Aubert, 1969). (The group is of marginal importance in Norway)	A. Climate is the most important commonality on the resource side. Good examples are many pensioners in northern USA who commute to Florida in winter and northern Europeans to the Mediterranean. B. The emigrant tourist represents a form of double settlement ('transhuman existence'), and has a demand for developing societal infrastructure and construction of houses and apartments.	– various types of accommodation abroad – rental accommodation – apartments – second homes	A. Norway lacks specific nature or culture-based resources that would generate residential tourism at a large scale. B. This type of tourism is particularly prominent within the western world but expands globally. Innovations within ICT provide preconditions for such an expansion.
THE SOCIALIZER AND SIGHTSEEING TOURIST is one who is looking to: – see the world in a rapid and comfortable way – seek the picturesque and exotic routes, 'to have been there' – impress in a social context, not least in relation to fellow countrymen	A. Traditionally, this form has been strongly oriented towards nature (visits grand scenic attractions). Therefore, the fjords in western Norway, the Lofoten mountains, Northern Cape, Spitsbergen, Greenland and the polar ice cap. B. This is a type of tourism that places very high demand on tourism suprastructure. Cruise ships, yachts and motorized vehicles are represented by this class.	– boat cruises – yacht trips – bus trips with overnights stays – dining to a high standard – hotels along typical sightseeing itineraries – a part of the charter trips by plane	A. Our resource potential has been nature-based and primarily in relation to boat cruises. Since customers can now be transported comfortably and relatively inexpensively by plane, our resource potential has therefore been weakened. B. The Pacific Ocean area and South America have emerged strongly. Internationally, this is regarded as an affluent and growing market. Norwegian producers (ship owners) have a large share of the global market, and they are market leaders because of high competence in product development and production.

Continued

Table 4.1. Continued.

Tourist type (needs)	A. Important resources B. Infra- and suprastructure	Product solution	A. The Norwegian resources situation B. The market
THE EXPERT/CONNOISSEUR is one who is looking to: – get cultural experiences in relation to the host community. Actively interested in local archaeology, geology, flora and fauna, cultural traditions and politics etc.	A. This group is resourceful and has often special interests within the fields of nature or culture. The group is oriented towards urban and cultural destinations, though certain activities such as hunting or photography may appeal to this group, too. B. The members of this group are recruited from upper classes and elites in industrialized countries. While activities in more rural environments have lower standards, in urban destinations this group looks for luxury and high-standard services.	Typical activities for this group include: – festivals such as Cannes, Bayreuth or Salzburg – city breaks in metropolitan areas – exclusive nature-based tourism such as trophy hunting or yachting	A. Norway has certain resources within this sector, but the utilization of rivers for hydropower has destroyed some opportunities for exclusive angling. Globalization has negatively affected the availability of local traditions. B. The market is economically resourceful and seldom affected by economic change. Globally, this market is increasing.
THE SUN WORSHIPPER is one who: – wishes to achieve well-being and enjoyment – desires for sun combined with other types of simple, sensual pleasure, such as cheap wine or spirits, exotic food and access to an enjoyable, relaxed existence for a few weeks (Aubert, 1969)	A. Strongly natural resource-based, focusing on sun and beach destinations during summer season and alpine resorts during winter B. The societal infrastructure must be well developed since it is catering for mass tourism. This means that the societal infrastructure and the tourism suprastructure must closely resemble the home environment.	– The majority of this tourist category is recruited from the middle classes. – The product is often standardized and produced in large volumes.	A. In an international context, Norway has relatively good conditions during the winter season in terms of natural resources (snow in abundance and long winters, suitable topography and ample access to terrain). Its two greatest 'handicaps' are that it is located in a periphery in terms of population, and lacks the exotic local society that makes a suitable backdrop. During the summer, resource conditions are rather poor. The temperature is too cool and the climate is not stable enough. B. Globally, there is expectation of continued growth. The Nordic region is not particularly suitable unless other, more suitable, production areas reach a saturation point.

THE ACTIVITY TOURIST is one who wishes to: – find activities that provide opportunity for physical exercise and other relevant leisure activities – experience activities appealing to various generations – find 'year-round activities'. Otherwise this tourist makes modest demands on the host area	A. The key resource here is knowledge, and this applies to both production development and production itself. B. The societal infrastructure should be well developed. Demands on the tourism-related suprastructure are accommodation and various services such as guiding etc.	– physical activities – trails for jogging – indoor ball games – swimming pool with a sauna – riding with an instructor – cycling paths – hiking trails – canoe trips with an instructor – tennis courts with instructors – ski trails – slalom runs	A. Norway has the natural resource conditions for developing a variety of products for this tourist group. In terms of knowledge, however, we remain somewhat weak in the area of both product development and production. B. During recent decades (from the beginning of the jogging trend in the USA in the 1960s), the market has grown rapidly and it is a relatively affluent market. Once again, there is a parallel to the cccurrence of health shops, increased sale of exercise and sport equipment, health studios etc.
THE UTILITY TOURIST is one who wishes to: – utilize portions of his holiday to gather provisions of fish, berries, wild game, firewood etc. for home use – live simply and inexpensively, partly due to a personal ecological conviction – an additional benefit for many is the collaboration of family members to help with tasks	A. Is exclusively based on naturally resources. In this respect is it rarely necessary that the resources be of 'outstanding quality'. The utility tourist is not there for trophy hunting, including fish. Harvesting natural resources is what is important. B. Requires some adaptation of tourism infrastructure and suprastructure. Specifically, this means a great deal of organizational work to make the resources available. Tools and other instruments need to be provided. Demands for comfort in tourism suprastructure.	– utility fishing in saltwater – utility fishing in freshwater – self-collected firewood – search for supplemental food supplies – provisions made for berry picking, mushroom hunting and other useful foods	A. The Norwegian resource situation, in an international context, is of a good standard offering both fresh water and seawater. In addition to this the natural resources have self-cut firewood and berry-picking to meet domestic demand. B. Both nationally and internationally, this is a significant market that is being developed. Less demanding to develop than the outdoor tourists and activity tourists.

- it is a method that establishes a relative measurement that can be used to distinguish between established tourist types. Landscape cannot be measured as having the same value for different types of tourists.

The sociological method, unlike the landscape architectural tradition, is not limited to only a few types of tourist. The sociological method also includes the somewhat more uncommon tourist types. Sociologically, they deviate from mass tourists and their properties are more heterogeneous. Still, when elements from these uncommon tourist types are encompassed in the mass phenomenon, as they often are, then they are particularly important (Aubert, 1969, p. 120). Aubert's work seems to support the current discussion on resource mapping, especially the need to address the sociological: if we wish to bring in economic wisdom pertaining to production, we must focus on niche productions and look for production conditions favourable to them, challenging the current easy-to-meet demand of the mass tourism market. Furthermore, the sociological method has been tested a number of times and has achieved its aim (see Aubert, p. 109 [bottom] and Annex 5). It primarily identifies several needs of the tourist and, by so doing, expands the spectrum of potential resources. Additionally, it incorporates an empirical method that is more precise than the indirect method that examines activities (the geographical method) and the landscape architectural traditions (the ethnocentric method) focusing on different values parameters: aesthetic, symbolic and identity value. However, when the resources have been identified, geographical terms and methods for describing and mapping will emerge as central.

Identifying resources, then, is an interdisciplinary challenge as Zimmermann emphasizes in his work:

> Both the analysis that preceded and the examples just cited make clear that the study of resources belongs in the field of the social sciences, but the synthesizing work of the social scientist must rest on the findings of both natural and applied scientists. The field of the social sciences is wide indeed. It embraces history, sociology, phases of anthropology, economics, government and related areas. The fields of the natural and applied sciences are even wider. The study of resources thus inevitably becomes the study of synthesis of knowledge...
>
> It is evident from the preceding treatise that no one group can, by itself, fully understand and appraise resources. These are joint responsibilities.
>
> The appraisal of resources proceeds from:
>
> **1.** the knowledge of facts of nature and culture to
> **2.** the determination of technical feasibility to
> **3.** the determination of profitability to
> **4.** the formulation of the grand strategy along socio-economic lines ...
>
> Two sets of interdependent criteria may be distinguished in the appraisal of resources:
>
> **1.** The State of the Wants, the expression of human needs, individual wants, social objectives, higher aspirations.
> **2.** The State of the Arts, as the summation of human capacities along lines of both technology and social organization.
>
> The constant flux in which these two sets of criteria exist connects itself to the resources to which the appraisal of these criteria applies. While the physical universe may be constant, while the planet earth may undergo few changes speedy and substantial enough to affect materially the destiny of man, the resources available to the human race are in constant flux, changing reflections of changing capacities and needs.
>
> (Zimmermann, 1972, pp. 17, 20)

The sociological method is highly relevant in assisting with the identification of tourism resources, through the theoretical paradigm such as Zimmermann proposes. It will primarily serve as the key to revealing what may constitute resources; beyond that, it will be necessary to activate knowledge from a large spectrum of different disciplines.

Notes

[1] In September 1992 I had the opportunity to speak daily with tourism researcher/professor emeritus Anthony S. Travis from Birmingham. At the time he was at Eastern Norway Research/Oppland University

College lecturing and providing research guidance. I took the opportunity to discuss Brian S. Duffield's assessment with him, which Professor Travis summed up by saying that during the last half of the 1960s it cannot be proved in the subject of geography that there was methodical progress of any significance with regard to surveying resources for leisure-related travel and tourism. To put it another way, Professor Travis confirmed that the paradigm that Duffield accounts for in his article of 1984 is still the standard that is applicable within the field of geography.

[2] Norwegian geographer Michael Jones likewise maintains that one must go via the consumers, as members of the culture, to gain access to a holistic conception of resources. In his work *Landscape as a Resource*, Michael Jones writes: 'The experience of beauty is emotionally charged, and for the individual, it will depend on specific experiences, associations and preferences. These are in turn influenced by culturally determined attitudes, values and notions of religious faith' (Jones, 1981, p. 1). Another important point to be remembered from Jones' work is that he also uses Zimmermann as a main reference when he explains *what resources are and what laws resources, as a phenomenon, are subject to.*

[3] Torbjørn Eggen, Head Curator of Røros Museum, has conducted a pilot study of the Røros plains under the theme 'Culturally Determined Use of Resources'. Here, the point of departure was to ascertain the use of resources by the Sami people (indigenous people in the area) and the use of the resources in the same area by permanent, ethnic Norwegian dwellers. The results of the pilot study conform entirely to what Duffield extracts from Zimmermann's theory of resources. To a great extent, resources are culturally determined and, as a consequence, what may be interesting resources for one population group may be insignificant or entirely uninteresting resources when considered for another group. Nordic Sami Institute, Kautokeino and Røros Museum, through the efforts of Head Curator Eggen, are currently working to continue the pilot study in a larger programme.

[4] For many years Søyland was president of the Norwegian Sea Fishing Association and a member of the board in two federations of sea-angling clubs.

[5] To avoid misunderstanding, the work referred to here was published for the first time in Germany, 1904–5, so that what Kelly refers to must be an English version printed in 1958.

[6] Point a), under the tourist categories 1–6, with the exception of the parentheses, is a repetition of Aubert's text describing tourist categories, pp. 8–9.

References

Aubert, V. (1969) *Det skjulte samfunn*. Pax, Oslo.

Bech, R. (1983) Det genfødte Polen: Landets folkesjæl er en attraktion i sig selv. *Politiken*. February. Kobenhavn, Denmark.

Bjønness, I.M. (1980) Animal husbandry and grazing: a conservation and management problem in Sagarmatha (Mt Everest) National Park, Nepal. *Norsk Geografisk Tidsskrift* 34, 59–76.

Blacksell, M. and Gilg, A.W. (1975) Landscape evaluation in practice: the case of south-east Devon. *Transactions of the Institute of British Geographers* 66, 135–140.

Blakar, R.M. (1989) *Forskning er ein livslang prosess*. Radet for samfunnsvitenskapelig forskning, Oslo.

Bruun, M. (1983) Vurdering av landskapskvalitet. Report. Institutt for landskapsarkitektur, Ås, Norway.

Chubb, M. and Chubb, H. (1981) *One Third of Our Time? An Introduction to Recreation Behaviour and Resources*. John Wiley & Sons, Chichester, UK.

Clawson, M., Burnell Held, R. and Stoddard, C.H. (1960) *Land for the Future*. Johns Hopkins University, Baltimore, Maryland.

Cohen, E. (1972) Toward a sociology of international tourism. *Social Research* 39(1), 164.

Cohen, E. (1974) Who is a tourist? A conceptual clarification. *The Sociological Review* 22(4), 527–555.

Coppock, J.T., Duffield, B.S. and Sewell, D. (1974) Classification and analysis of recreation resources. In: Lavery, P. (ed.) *Recreational Geography*. Wiley, Chichester, UK, pp. 231–258.

Coppock, J.T. and Duffield, B.S. (1975) *Recreation in the Countryside: A Spatial Analysis*. Macmillan, London.

Crofts, R. (1975) The landscape component approach to landscape evaluation. *Transactions of the Institute of British Geographers* 66, 124–129.

Crofts, R.S. and Cooke, R.U. (1974) *Landscape Evaluation: A Comparison of Techniques*. University College London.

Duffield, B.S. (1984) The study of tourism in Britain: a geographical perspective. *GeoJournal* 9(1), 27–35.
Fredriksen, D. and Møller Sørensen, J. (1977) Ungdom, utdannelse og fritid. Munksgaard, København, Denmark.
Geelmuyden, A.K. and Berg, T. (1986) *Vassdragsreguleringer og landskap: En oversikt over kunnskapsnivået.* Ås, Norway.
Gilg, A.W. (1975) The objectivity of Linton type methods of assessing scenery as a natural resource. *Regional Studies* 9(2), 181–191.
Hellevik, O. (1991) *Forskningsmetode i sosiologi og statsvitenskap*. Universitetsforlaget, Oslo.
Jones, M. (1981) Landskapet som ressurs. *Dugnad* 7(1), 1–15.
Kaspar, C. (1975) *Die Fremdenverkehrslehre im Grundriss*. Haupt, Bern.
Kelly, J.R. (1987) *Freedom to Be: A New Sociology of Leisure*. Macmillan, New York.
Kruschke, E. (1974) *Consequences of Failure*. Sage, London.
Langdalen, E. and Stav, O. (1974) *Jeløya. Analyse av verneverdier i naturmiljø og kulturlandskap*. Institutt for Jordskifte og Eiendomsutforming, Ås, Norway.
Lykkja, H. (1991) *Reiseliv og areal: Vedlegg II: Meir om metoden*. Norsk Institutt for Jord- og Skogkartlegging, Ås, Norway.
Lykkja, H., Øyen, B.H. and Rekdal, Y. (1990) *Turist-Norge: Reiseliv og areal*. Norsk institutt for Jord-og Skogkartlegging, Ås, Norway.
Meistad, A. (ed.) (1974) *Seljestadomradet, Odda kommune: Idékonkurranse om utforming av Seljestadområdet for friluftsliv- og ferieaktiviteter*. Norske Arkitekters Landsforbund, Oslo.
Müller, H.R. (1992) Ecological product declaration rather than 'green' symbol schemes. *Revue de Tourisme* 47(3), 7–10.
NOU (1974) Viltstellet, jakt og fangst. 21. Miljøverndepartementet, Oslo.
Parker, S.R. (1965) Work and non-work in three occupations. *The Sociological Review* 13(1), 65–75.
Penning-Rowsell, E. (1973) *Alternative Approaches to Landscape: Appraisal and Evaluation*. Middlesex Polytechnic, Enfield, UK.
SOU (1971) Mark och vatten. Statens Offentliga Utredningar, 1971:75. Civildepartementet, Stockholm.
Søyland, W. (1985) Fritid – strabaser og store opplevelser: Om fiske fra land og båt langs norske kysten. In: Svalastog, S. (ed.) *Turisme ved kysten*. Oppland distriktshegskole, Lillehammer, Norway.
Statistics Norway (1983) Sosialt utsyn 1983 (Statistiske analyser). Statistisk Sentralbyra, Oslo-Kongsvinger, Norway.
Statistics Norway (1989) Sosialt utsyn 1989 (Sosiale og økonomiske studier). Statistisk Sentralbyrå, Oslo-Kongsvinger, Norway.
Svendsen, H.T. (1990) *Reiseliv og areal. Vedlegg II: Landskaps-kartlegging – Regioninndeling – klassifisering*. Norsk Institutt for Jord- og Skogkartlegging, Ås, Norway.
Travis, A.S. (1982) Physical impacts trends affecting tourism: managing the environmental and cultural impacts of tourism and leisure development. *Tourism Management* 3(4), 256–262.
US Department of Agriculture (1974) *The Visual Management System*. Agriculture Handbook 462. USDA, Washington, DC.
US Forest Service (1979) *Visual Character Types*. US Forest Service. Alaska Region, Division of Recreation, Anchorage, Alaska.
Zimmermann, E.W. (1951) *World Resources and Industries*. Harper & Row, New York.

5 On the Production Process and International Competitiveness

Sondre Svalastog

Introduction

In terms of production, tourism is nothing special. The only special production-related aspect in this sector is that a striking number of politicians, bureaucrats, tourism professionals and institutions act as if the sector should be cut loose from all the ordinary theoretical laws governing production. In my view, this is a result of the sector's belated inclusion in the sphere of science. My hypothesis is that the sector, from a production perspective, is roughly as predictable as any other sector. The local and global laws governing localization and division of production are no less valid in this sector than they are in any other production cycle.

It is a labour-intensive sector, and within production we are therefore able to see a very clear centre-periphery dimension. Significant parts of standard production that are vulnerable to competition (parts of resource-based tourism) are localized in low-cost countries with climate advantages, while for high-cost countries such as Norway and the Nordic countries, knowledge-intensive productions, which show specialisms, have far greater potentials. If knowledge acquisition lags behind – such as has been the case in Norway – the following occur: the ability to generate competitive products begins to wane and loses volume and share in the market; and the inland centre-periphery dimension in the business sector is bolstered because the tourism production that is vulnerable to competition, much more prevalent in the periphery, does not share the same opportunities to raise prices as the protected production in the cities and urban areas.

The Harvard study, *Competitiveness of Nations*, documents the following four basic factors that can be applied to our context:

> The country's rise in costs incidentally justifies or underpins the growth of internationally outstanding enterprises, but there are other factors that truly drive competitive power. (Strandell, 1988, p.14)
>
> Direct government influence on the industry's competitiveness is obvious. Generally, one might say that government policy will be unsuccessful if the other prerequisites for growth of competitive enterprises are not present. (Strandell, 1988, p. 19)
>
> Competitiveness is not achieved or inherited without something being created through continual efforts. Special factors that require comprehensive investments and are difficult to imitate – technological advances and specialized education are the more sustainable competitive advantages... (Strandell, 1988, p. 89)
>
> Through studying subfields and enterprises, one obtains a richer and more nuanced image of what stimulates, and conversely what counteracts international progress in a national perspective. In this perspective, entire countries are regarded as competitors. Rather than a unified set-up, countries are instead seen as more or less competitive subgroups. (Strandell, 1988, p. 14)

 Tourism as a Resource-based Industry: Based on the work of Sondre Svalastog (eds A.L. Svalastog, D.K. Müller and I. Jenkins)
DOI: 10.1079/9781800621480.0005

However, it is unrealistic to guarantee international competitiveness solely by focusing vast resources into one particular sector, i.e. tourism.

The findings of the Harvard study confirm basic elements in the theoretical notion of production which I have applied in my fieldwork. International competitiveness in a sub-field can be sustained and even further developed quite independently of wage and cost levels in the producing country. What we face, however, are the following two governing production principles:

1. The know-how level must follow the wage/price increase in a country if the conditions for international competitiveness are to be sustained.
2. If the real competence, in an international sense, of a nation is raised more quickly than the wage/price level, the nation's competitiveness will also rise in the area in question; and *vice versa*, if the level of competence in a nation is in decline it will also decline internationally.

The following two clarifications, however, must be linked with the governing principles stated above. First, we must equip ourselves with a pair of terms that enable us to describe the development of knowledge that takes place (more generally, or within a sector) within a country. This can be in contrast to global knowledge development. I would term the first factor *knowledge development in a subjective sense*, and the other as *real knowledge development*. That is to say *subjective knowledge* is limited by the country's spatial boundaries whereas *real knowledge* is exchanged on an international level, which does not have these geographical restrictions. In other words, the international knowledge development is more rapid than the national development. In actuality, there is a decline in know-how in the country in question. Thus, it is the international knowledge that counts, not the knowledge available within a country.

Secondly, as was done in the Harvard study, I will also distinguish between general competence within a sector and expertise/know-how associated with parts of the business sector (Strandell, 1988, p. 14). General competence in the sector seldom creates international competitive advantages.

The wage and cost level, viewed in isolation, is a *framework condition* from which one needs to know how to understand the correct outcomes. The most important and decisive single factor for obtaining an international competitive advantage is the knowledge factor and, in this respect, it is as much knowledge for *product development* (i.e. competence, theoretical and practical; creativity; and initiative) as it is knowledge for *production*. The public sector cannot play any creative, direct role in production and product development. The politicians and political bodies, however, may, indirectly, over time, play a crucial role in a business context by making provisions for *real and adequate* knowledge production at our educational and research institutions.

Production and Knowledge

To produce, in an economic sense, means to develop an individual resource (input factor) or a cluster of individual and/or partial products so that:

- what comes out of the production process will be motivating for the market, i.e. that a group of people (market segment) assesses the end-result as a good response to their need for the product or service that they may want; and
- the production process can be evaluated on the basis that the outcome is higher than the value of the individual factors that were put into it. The more 'value added' that results from the production process, the greater the market competitiveness of the product.

'Knowledge is incomparably the most important of human resources', Mitchell (1941) states, and acceptance of this assertion will be used in the analysis that follows (the reference merely underscores and supports conditions and premises I have already established). Furthermore, one maintains that a product's properties are a result of:

1. the individual input factors; and
2. the real competence invested in product development and production.

These connections can be illustrated in Figs 5.1–5.3.

Figure 5.1 provides a schematic image of the phases in the production process. In the course of this, the input factors are processed so that they

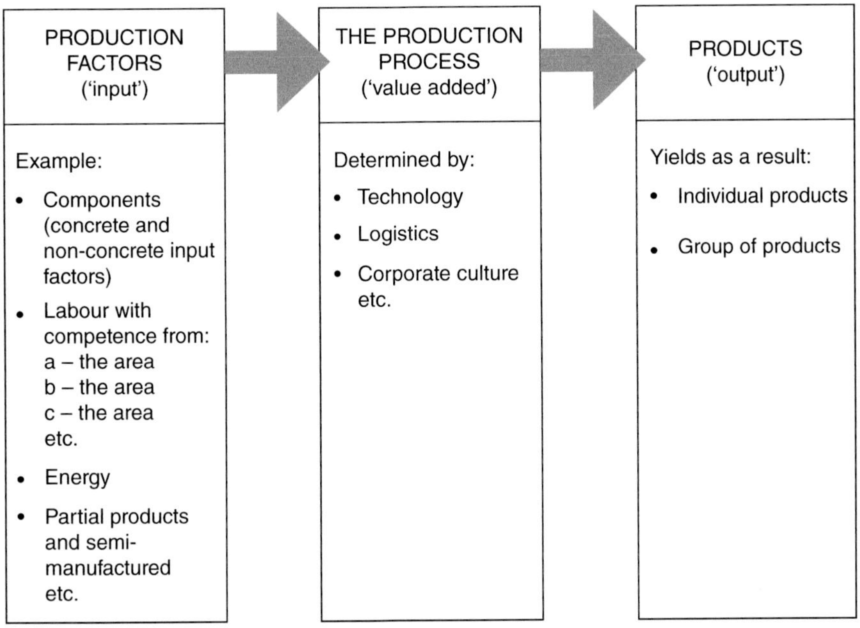

Fig. 5.1. Phases in production from a technical perspective. (Adapted from Svalastog, 1986, p. 157)

change the properties[1] or form[2] and are therefore in demand in a larger or smaller market.

The quality of the input factors, as well as the production process itself, is of crucial importance for the quality of the 'output'. First-rate raw materials cannot ensure first-rate products if there are weaknesses in the production process itself. On the other hand, it is equally difficult to create first-rate products from raw materials that are not of a high quality. For this reason, a close connection exists in the tourism sector between our ability to conduct good professional resource analyses and the products and product combinations we put on the market.

In Fig. 5.1, and in comments, I have pointed out the basic context with which the production process is technically associated, but the previous stage (the product development stage) and the subsequent stage (the marketing stage) are equally important for us to understand.

Before production can be initiated, the production process must be conceived. Thoughts that establish objectives are indeed prerequisites for a conscious process and it is thoughts and ambitions (objectives) that create value. Prior to the existence of concrete products and provisions made for production, ideas promote an interplay of factors leading up to the product; it is ideas that trigger notions pertaining to the functions and needs the products will satisfy. Therefore, the key aspects of our next step will be to visualize the meaning of the stage prior to the production process itself, pointing out the insights from which we derive the creative process.

The stage of the production process itself can be schematically linked to this, as shown in Fig. 5.2.

As we see from Fig. 5.2, know-how must be a key factor. It determines our *ability to develop products* and strongly influences the production process itself and thereby the *quality of the end-product*. Some of the main points in Figs 5.1 and 5.2 can be illustrated through the following examples.

Some years ago, the Soviet Union ordered an entire car factory from the industrial group Labyrinthal Italian Automobile Torino (the Italian company that makes Fiat cars and lorries). The turnkey factory was delivered to the Soviet Union and was intended to begin manufacturing copies of Fiats. In this example, there was common know-how behind the product concept. The input factors in their unprocessed form were qualitatively alike, while the manufacturing

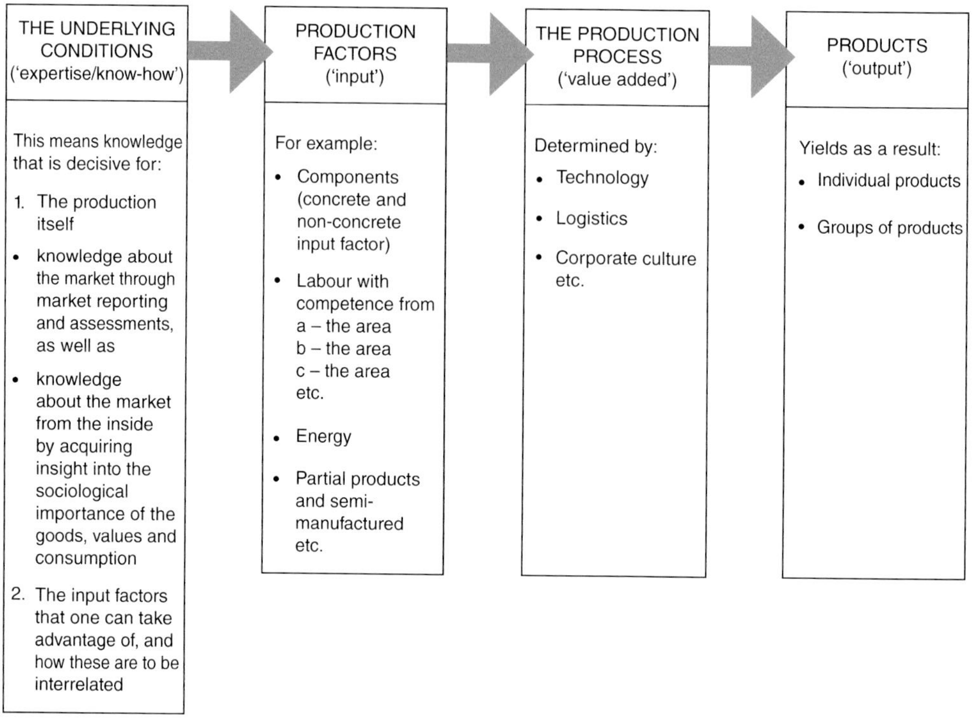

Fig. 5.2. Product development and production in a technical sense.

technology was the same. The results, however, documented with blinding clarity that the Russian Lada, as it was called, and the Italian-manufactured Fiat were qualitatively two different products. The differences must have arisen during the production process itself. Essentially, this can be attributed to different levels of know-how among factory workers, differences in corporate culture and differences in organization between the Russian and Italian modes of production. The fact that the Russian Lada was sold in significant numbers in Norway is due to price differences.

The Per Spook fashion house in Paris sells a complete package to the Norwegian market for people who want to knit their own fashionably chic, warm sports jumper (the package contains knitting instructions, yarn and the Per Spook emblem). The product concept has such market appeal that the package is sold at a price that is 50% above what the input factors (yarn and standard instructions) would have cost separately. The fashion concept is marketable because the sports jumper has a higher abstract value than the utility of the item (keeping the body warm) and the product's image has a very high 'value added' competence level.

Given its relatively small size, Norway has a sizeable IT industry. This knowledge-intensive industry is a direct result of a long-term accrual of knowledge by some of our most important research environments, primarily the Norwegian Defence Research Establishment, Kjeller, the Norwegian Institute of Technology and the natural sciences and mathematics environment of the University of Oslo.

Usually, innovations do not happen by chance; innovations are normally the result of systematic work over a long period. That is why Michael E. Porter, for educational purposes, has included a separate box in his model representing unplanned and random innovation. The box is placed outside the main model as a counterpart to a systematic and intended accumulation of knowledge (Porter, 1990, p. 15). If a commodity vulnerable to competition (as is a significant portion of resource-based tourism) is to be produced in a high-cost country such as Norway, the knowledge input must be very high. It is knowledge that gives the production process its value-added dimension and resource-based tourism is no exception to this rule.

In production theory, a distinction is made between production in a technical sense and production in an economic sense. The technical aspect is linked with the fact that output has another identity than the goods and services that went into production. However, one must remain aware of the fact that 'the transformation we call production in the technical sense does not necessarily change the very material properties of the product itself. It can often be merely a movement or sorting or storage – "a movement in time"' (Frisch, 1962, p. 15). The term 'production', in an economic sense, on the other hand, is linked to the effort to create a product that has an output value higher than the input (Frisch, 1962, p. 20). This concept is what gives meaning to the phrase 'a good is not produced before it is sold'. To illustrate the difference between production in the technical sense and production in the economic sense, we must adapt the production diagram as shown in Fig. 5.3.

Since marketing is not included in the scope of this work, it will not be discussed further than what is necessary to ascertain the basic contexts for what I call the *productive system* (Fig. 5.3).

First, marketing can indeed be regarded as an independent and key component in the production process in an economic sense. This is because the major challenge is to bring products to markets in which they are perceived as meeting the current need for goods and services. Knowledge-intensive production demands more from the marketing function. Market niches must be defined, found and processed in a professional manner. A goal-oriented market communication requires both *product knowledge* and highly developed *communication skills.*

Secondly, there is a close dependency between product development and marketing. If one neglects the product development phase in terms of resources, there will be few market-oriented products to promote. On the other hand, the marketing phase is key for locating market niches, developing them and establishing good connections between producers and the market.

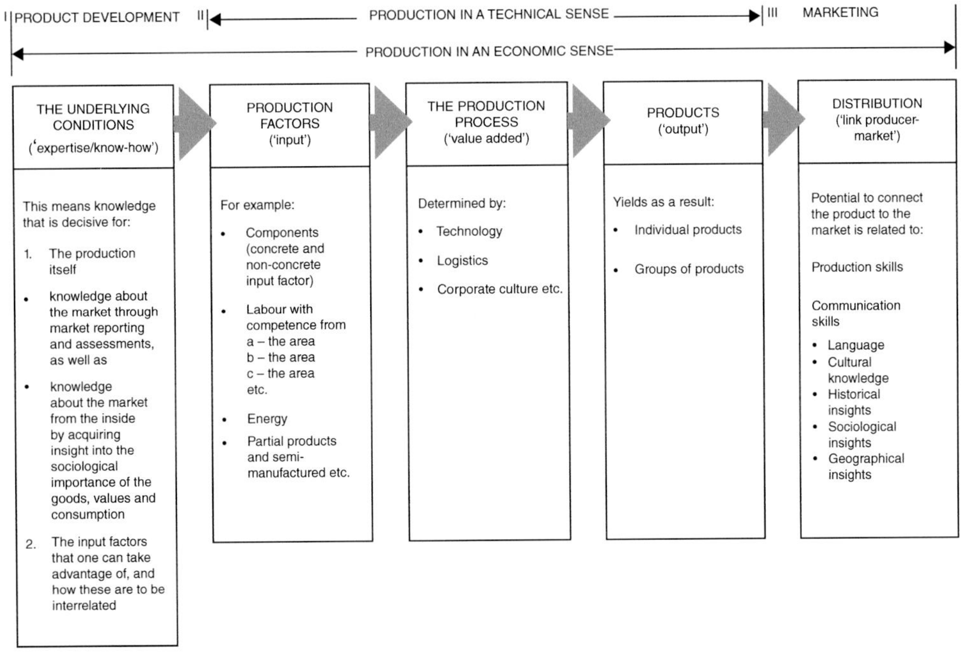

Fig. 5.3. Production in a technical sense and production in an economic sense.

Thirdly, close collaborative relationships must be developed between the product development function and the marketing phase. The marketing stage is usually aware of the mode/demand in various markets and should therefore develop good channels for information flow, exchange of experiences and channelling knowledge to the production stage (Phase I: Product Development in Fig. 5.3).

Norwegian Tourism During the Post-war Period

During the post-war period (1945–1990), Norway has changed from being a low-cost country to a high-cost country. An alteration in the conditions for production of this kind places great challenges on the tourism sector. In the following discussion, I shall devote some space to the amplification of the main characteristics from this period and to identify explanatory variables behind this development phase.

The first indicator we shall look at is how the total tourist volume for the group of 'hotels and other overnight facilities' developed during the period 1950–1990. In Fig. 5.4a, we find a curve representing total volume and scope of foreign visits. In Fig. 5.4b we show the absolute movement in foreign tourism in order to remove the 'equalizing effect' that is contained in the main curve in Fig. 5.4a.

To obtain an understanding of how key production conditions have changed throughout the period, I have chosen to divide the period into the following three phases:

- The first phase (1950–1960) includes the period of reconstruction and commercial establishment in conformity with the 'National plan for the reconstruction, modernization and new building of hotels, inns, tourist stations etc.' (Svalastog, 1988, pp. 104–105). During this phase, old and run-down production facilities were modernized and a large number of new small-size companies were established.
- The second phase (1961–1975) represents the time interval when the sector primarily embarked on economy of scale, as the production companies were provided with a significantly more favourable relationship between administration and production. A product renewal also took place since this was the era of building indoor swimming-pool facilities. Many new ski lifts were also added.
- The third phase began in the mid-1970s, and the developments were not really positioned in the appropriate areas of tourism. This phase is characterized by a rapid development of studies in tourism at our regional university colleges. Parallel to this, the regional research foundations began to become strongly involved in fact-finding

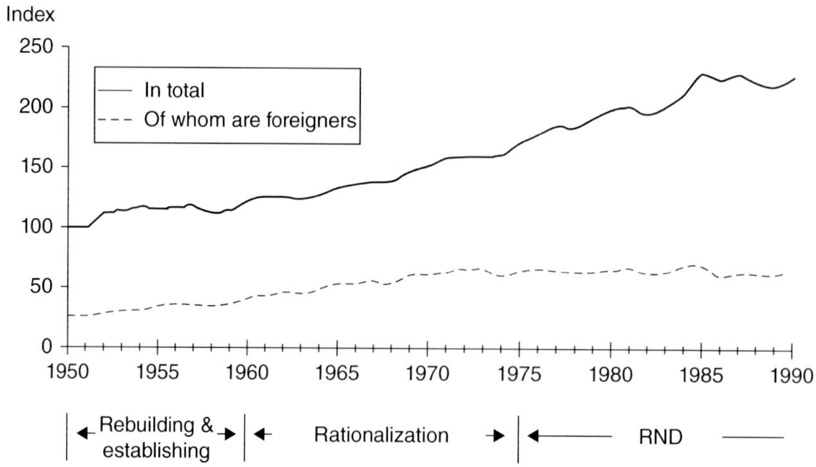

Fig. 5.4a. Overnight stays in Norwegian hotels and inns, 1950–1990 (Index 1950 = 100).

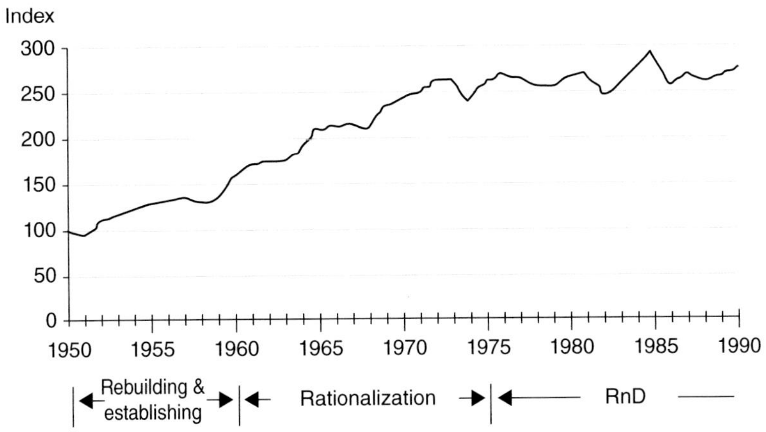

Fig. 5.4b. Foreigners' overnight stays in Norwegian hotels and inns, 1950–1990 (Index 1950 = 100).

reports on tourism as a discipline. (For more information on structure and scope, see NLVF, 1988, pp. 33–35).

One of the most important conclusions we can draw from Figs 5.4a and 5.4b is the following:

- Total overnight tourism to Norway, including both the sheltered part (profession-based tourism and Norwegian travellers' short-term, leisure-oriented trips) and the part that is vulnerable to competition (Norwegians' and foreigners' holiday-related demand), shows very stable growth through the entire period. Within the statistics there is a misleading deviation for the period 1984–87 when many hotels and inns were used as long-term housing for asylum seekers and refugees (therefore, this particular datum, 1988, was not included in the overnight stay statistics).
- Foreign tourist volumes amounted to 26% of the total in 1950 (benchmark year) and grew to 35% of the total in 1961; it amounted to *73% of the growth* in the phase before reconstruction and development.
- At the close of 1975, foreign tourism accounted for 38% of the total in the country. Foreign tourists accounted for 47% of the *total growth* in traffic during the *reorganization phase.*
- At the end of the period (1990), foreign tourists accounted for 29% of the total traffic and of the total growth in Norwegian overnight facilities during the period 1975–1990; foreign tourists *barely* contributed *1% of the growth.*

So far, the RND phase has not been successful in terms of production. Over a 15-year period, there has been no growth in foreign tourists in Norway. The growth in Norwegian overnight facilities has been buoyed by the growth in the protected and non-competitive portion of the market.

Figure 5.3 is an interesting indicator when it is used together with information we have from Figs 5.4a and 5.4b. Although the travel currency information includes both expenses attributed to the protected demand and the demand in the part of the market that is vulnerable to competition, Fig. 5.5 is primarily an indicator of the way the industry's international competitiveness has developed.

At the beginning point (1950), imports were somewhat larger than exports (186 M Norwegian kroner (NOK) in contrast to 125 M, respectively, measured in 1950s NOK value). At the start of the reorganization period (1960), exports and imports of tourism services were in roughly the same proportion (251 M NOK, in 1950 kroner values, both ways).

Export development during the reorganization period (1961–1975) was beneficial for Norwegian tourism up until the oil crisis in 1974. From that point, Norway experienced a rapidly rising wage and price level, compared with the nations that were its most important partners in trade (NOU, 1988, pp. 21, 30–34).

This development continues uninterrupted over the period of a decade and places Norway

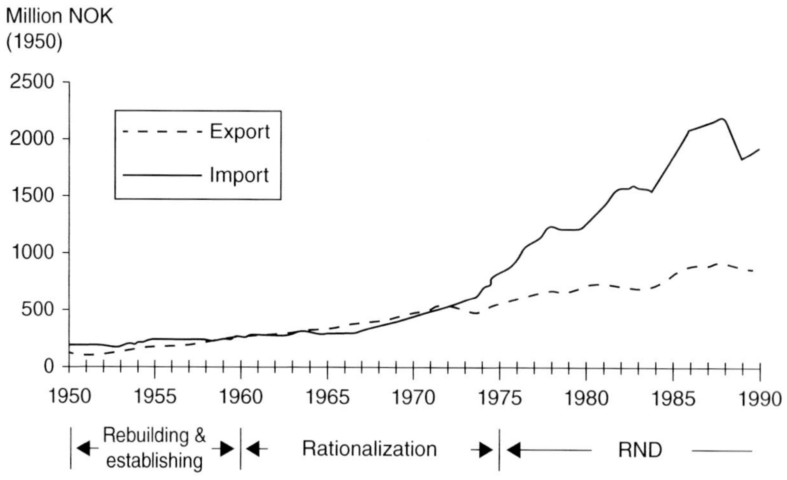

Fig. 5.5. Travel currency revenues (export) and fees (import) (converted to 1950 year's NOK).

as a prominent, high-cost nation in an international context. By the close of the reorganization period, international competition lightens somewhat, so that Norwegian imports were 2.25 times larger in 1990 than exports (respectively, 1895 M NOK compared with 836 M NOK in 1950 NOK).

The graph for the sector's international competitiveness (Fig. 5.5) tells us that the old product concepts, along with modernizing of the production apparatus and economy of scale sustained and partially strengthened Norwegian tourism's international competitiveness during the period 1950–1971. In 1971, export of tourism services amounted to 478 M NOK in 1950 kroner, while imports amounted to 465 M NOK in 1950 kroner). This is the last year when exports of tourism services were greater than imports. During the period 1972–1990, there was a dramatic reduction in international competitiveness in Norwegian tourism.

At the same time as Norwegian tourism lost international competitiveness, there were significant changes in both the winter and summer tourism products. The winter product altered its composition through a large investment in alpine facilities (in 1971, a total of 145 ski lifts and drag lifts; in 1975, 155 facilities and in 1990, 635 facilities). What is notable here, in a theoretical sense, is that the alpine product is rarely anything more than a partial product that must be supported in combination with several other product components. The Norwegian alpine product, in terms of quality, ranks below that of central Europe, because central European facilities are more comprehensive, often linked together so that they provide regular trip options and are competitively priced. The alpine product, in isolation, is a *mature, standard product* which may be, in a theoretical sense, a sustainable product innovation in a high-cost country like Norway. The investment in the alpine product in this period should be labelled as a *necessary product expansion*, but the preconditions did not facilitate *sufficient product development*.

However, when viewed from competition between internal Norwegian holiday venues, the investment in alpine facilities has resulted in the outcomes shown in the series of figures 5.6a–c.

The Norwegian alpine areas covered by Figs 5.6a–c during the five-year period 1984–1988 have seen a significant increase in tourists, while the cross-country area has been in decline. Moreover, evidence suggests that the alpine areas are those that come out best in terms of tourism during the summer. This is possibly because of the repeat-purchasing effect on the part of a satisfied winter customer; but most probably this is as much an indication of the fact that the almost pristine natural product that is offered by the alpine areas during the summer is regarded by the market as equivalent to what the country's traditional trekking areas have to offer.

During the 1980s, the Norwegian summer product acquired a new product component in the form of amusement parks, play lands and

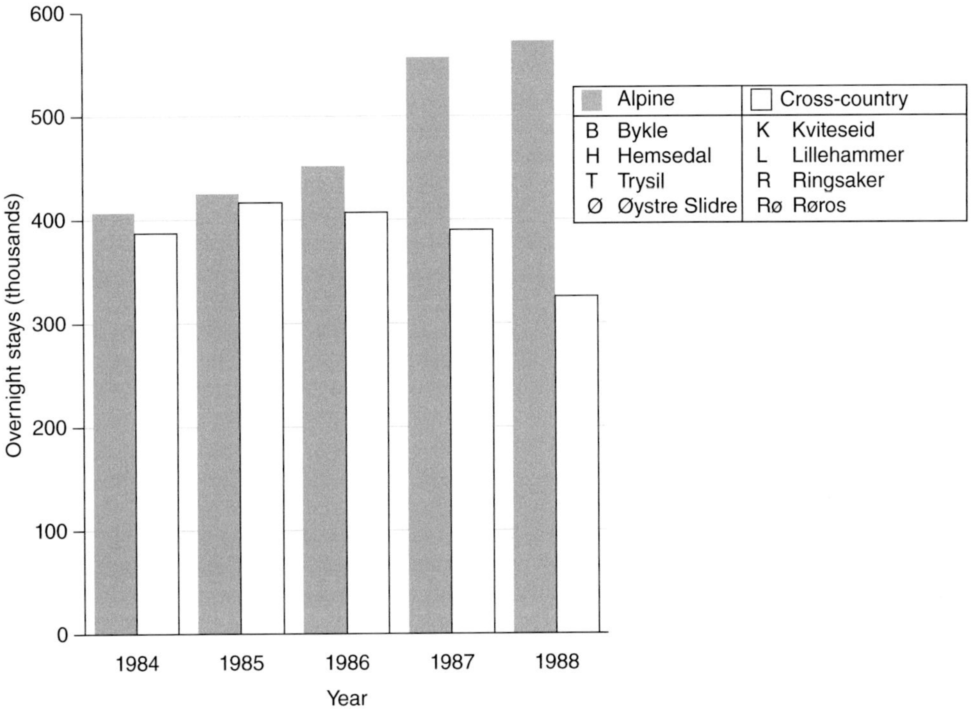

Fig. 5.6a. All overnight stays in hotels and other overnight facilities for four alpine and four cross-country resorts (figures in thousands). (From Statistics Norway, Kongsvinger)

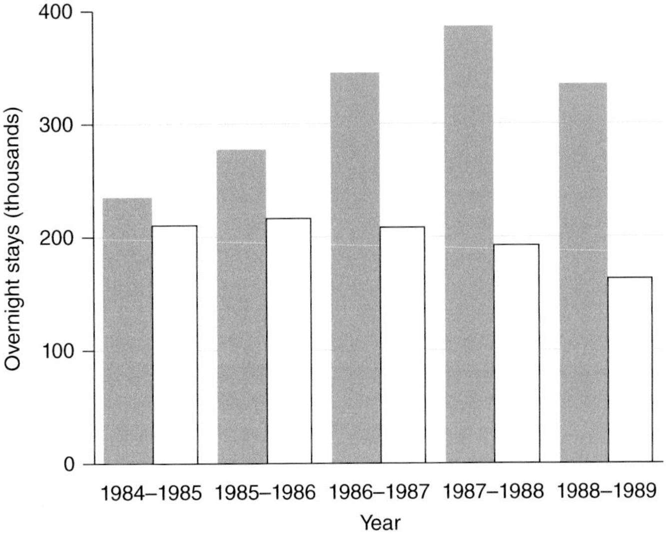

Fig. 5.6b. Winter seasons: all overnight stays in hotels and other overnight facilities for four alpine and four cross-country resorts (figures in thousands). (From Statistics Norway, Kongsvinger)

fantasy landscapes. As attractions, these are synthetically created, not site-specific attractions and we therefore find parallels to these in many other countries. Because of Norway's peripheral geographical location, these partial products can hardly have any substantial

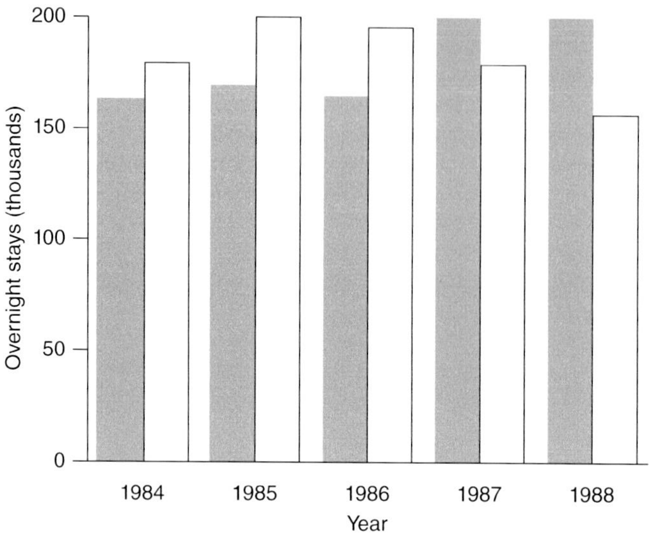

Fig. 5.6c. Summer seasons: all overnight stays in hotels and other overnight facilities for four alpine and four cross-country resorts (figures in thousands). (From Statistics Norway, Kongsvinger)

impact on the international competitiveness of the sector.

The product renewals/innovations that occurred in Norwegian tourism during the 1970s and 1980s are not, in the sense of production theory, of such a nature that they can sustain the industry's international competitiveness, or strengthen it.

Necessary Product Renewals Are Not Necessarily Sufficient Product Renewals

The terms *necessary product renewals* and *sufficient product renewals* derive from economics and reflect the simple fact that changes must be made in a product or a cluster of products if they are to have a future place in the market in which they have been incorporated. This does not mean, however, that the product will undergo a change in its basic nature. Standard products will continue to remain standard, even though necessary renewals take place in parts of the product. The standard Norwegian winter tourist product will continue to be an ordinary, standard winter product despite the fact that it may gradually change from being a cross-country ski product to one of standard alpine facilities, which dominate the market. All hotel owners and potential investors in winter sports destinations within industrialized countries have access to technology and equipment that are well developed and require little know-how to use.

In principle, one can say the same, for example, about various types of indoor water parks. Technology required for construction of this kind of production facility is universally known and production does not require a unique competence.

The main point is that product renewal for ordinary products and product clusters is necessary but will not transform them into special products needing higher levels of competence. They will continue to remain within the same area in the product cycle irrespective of changes.

Sufficient product renewal means a renewal of the product that entails a turnover that is profitable. In a qualitative sense, this means that managing the product will require more competence and result in a special product. It will also change character from a *mass product* towards the status of becoming a *special product*. In Fig. 5.7, the competence-intensive, special and niche products will lie in the left half of the curve, whereas the standard products will lie in the right half of the curve.

Using production theory, I referred earlier to product renewals that have taken place in Norwegian tourism during the past 10–15 years, inferring that it may indeed have been *necessary*,

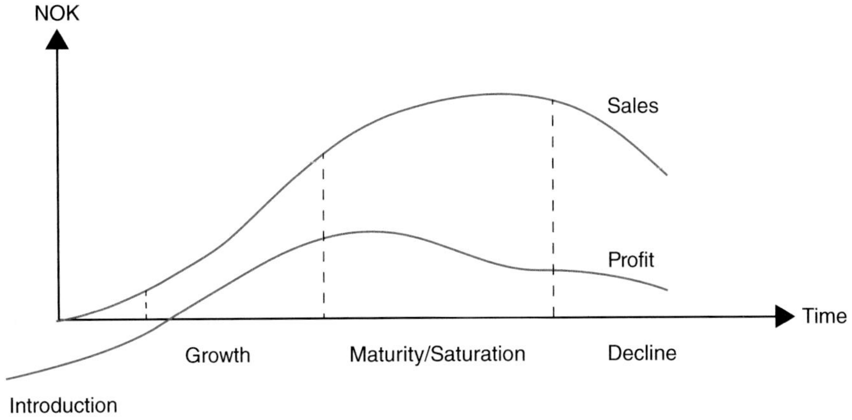

Fig. 5.7. The general life-cycle curve.

but that, economically, it has not been *sufficient.* Quantitative data seems to support this assertion as during the period 1986–1991, which is the period for which we have statistics from SSB, the total tourism for Norwegian hotels has increased. The equity share, however, year by year and for the entire period, has declined in the industrial classification of business sector 63: Hotel and Restaurant (see Table 5.1).

A sufficient product renewal is primarily dependent on competence across a broad spectrum.

> Brainpower has always been an essential asset… But it has never before been so important for business. Every company depends increasingly on knowledge – patents, processes, management skills, technologies, information about customers and suppliers and old-fashioned experience. Added together, this knowledge is intellectual capital.
>
> (Stewart, 1991)

Summary

We know a great deal, theoretically, about how international competitiveness in a sector can be sustained and developed further. What is difficult in practice is to conform to what we know.

Gaps in knowledge and a lack of specific action to rectify this failed to improve international competitiveness in the tourism sector during the period 1972–1990. The following three factors explain some of the reasons.

First, there are basic weaknesses in the way the research and education system has been developed from 1972 until the present time. Characteristically, the resources have been distributed to *many institutions*. The study programmes that have been developed are general introductions to the *entire field of tourism*, the majority of the studies are based on *business and administrative subjects* and the elementary introductions to the discipline are not commensurate with international research developments.

The many institutions are small in size and therefore have limited resources to provide for specialization (advanced study). Specialization is necessary if one is to have the opportunity to establish an adequate knowledge base for knowledge-intensive specialisms. The Harvard study entitled *Competitiveness of Nations* sufficiently documents the fact that international competitiveness is linked with special competence in parts of a sector, regardless of general insight into an entire sector. The Norwegian study programmes of 1972–1990 lacked specialized topic areas and offered general knowledge in tourism to satisfy students' curiosity.

During the 1960s, the Norwegian Regional Development Fund released materials documenting that most of the Norwegian tourism enterprises that had gone bankrupt or suffered serious economic problems, had failed to adequately manage the economic aspects of their businesses. There is no reason to doubt that this failure was the weak link and the direct cause of economic difficulties in the majority of the

Table 5.1. Equity share in Norwegian hotel business. (From statistical weekly pamphlet no. 27, 1992, Statistics Norway, Oslo/Kongsvinger)

Year	Equity share
1986	10.3%
1987	12.3%
1988	9.6%
1989	5.2%
1990	3.0%
1991	2.4%

problem enterprises. However, in terms of education, this realization led to erroneous conclusions where courses in economics were expected to provide sufficient and significant management tools and financial control.

The last, and equally conspicuous, weakness associated with the Norwegian education system in the field of tourism is that it comprises short-term studies without connecting to international research. The politicians and bureaucrats in the field of tourism have developed an educational and research structure that, when viewed subjectively, has probably contributed towards developing knowledge in the field of tourism. However, in its present form the system is not up to par in an international context.

The second main factor limiting the level of international competitiveness in Norwegian tourism is the randomness of products and production places (areas of commitment). The portion of Norwegian tourism that is vulnerable to competition is resource-based production, where *knowledge* and *resources* must be made mutually dependent variables and given a high influence.

The third main area where the lack of competence is severe is in the connection between producer/market (cf. the marketing box at far right in Fig. 5.3). The marketing function must be able to act skilfully in terms of both the market and product developers. If this is to occur, the marketing function must possess both a higher level of competence and adequate specialization.

The main conclusion from this chapter is: tourism is production and this production is subject to the same theoretical laws of production as all other production fields. If a high-cost nation such as Norway is to develop international competitiveness in parts of the tourism sector (with the exception of Norwegian cruise activity, which is a successful exemplar of this discourse), the industry must change its focus to developing knowledge and competences that can capture international markets with international products.

Notes

[1] Physically speaking, the salmon river did not change character when it became an attractive tourist product, but there was nevertheless a production at play when specific agreements were entered into with property owners, capital was advanced and the product was put on the market etc.

[2] A skiing weekend at a winter sports venue may comprise, among other things, partial products such as slalom during the day and *après ski* during the evening and night-time. During production of the partial slalom product, the key input factors change the character. Vegetation is removed from large surfaces; the topography is reworked and the local climate is radically changed since new drainage areas for cold air may be established and winter precipitation in the form of ice and snow are artificially made much more expansive on the ski trails and in the target areas, in order to prolong the winter season. With this kind of production, there is a significant physical change of input factors up until the created product is useable.

References

Frisch, R. (1962) *Innledning til produksjonsteorien* (9th edn). Universitetsforlaget, Oslo.

Mitchell, W.C. (1941) Conservation, liberty, and economics. In: *The Foundations of Conservation Education*. National Wildlife Federation, New York. [Cited in Zimmermann (1972)].

NLVF (1988) Om behovet for forskning ogutvikling innenfor bygdeturisme. NLVF-utredning nr. 146. Norges landbruks-vitenskapelige forskningsrad, Oslo.

NOU (1988) Norsk økonomi i forandring. Perspektiver for nasjonalformue og økonomisk politikk i 1990-årene, i Norges offentlige utredninger. Universitetsforlaget, Oslo

Porter, M.E. (1990) *The Competitive Advantage of Nations*. MacMillan, London.

Stewart, T. (1991) Brainpower: how intellectual capital is becoming America's most valuable assets. *Fortune* 3, 44–60.

Strandell, A.-C. (ed.) (1988) *Sverige – Drivhus för internationell konkurrenskraft?* Allmänna förlaget, Stockholm.

Svalastog, S. (1986) Reiseliv och noen produksjonsøkonomiska lovmessigheter. Paper presented at the Nordiskt forskarsymposium on tourism, 20–21 March. Göteborg, Sweden.

Svalastog, S. (1988) Tourism in Norway's rural mountain districts twenty-five years after the Mountain Planning Team's report. *Norsk Geografisk Tidsskrift* 42(2/3), 103–120.

Zimmermann, E.W. (1972) *World Resources and Industries*. Harper & Row, New York.

6 National Parks, Protected Areas and Tourism Labour Markets in Arctic Sweden

Dieter K. Müller

Introduction

Tourism development can act as a remedy to create growth in rural and peripheral areas, particularly in times of economic decline (Jenkins *et al.*, 1998; Baum, 1999; Fleischer and Felsenstein, 2000; Hall and Boyd, 2005; Hall, 2007; Jóhannesson and Huijbens, 2010). It has also often been seen as a way to change demographical structures, not least by attracting young people to remain in – or to enter – rural and peripheral areas (Paniagua, 2002; Müller, 2006; Schmallegger *et al.* 2011). In this context particularly, national parks and protected areas might be important resources with an ability to lure people over great distances to even remote places in order to enjoy pristine nature and wilderness (Moisey, 2002; Buckley, 2004; Fredman *et al.*, 2007; Hammer, 2007; Wall *et al.*, 2007). Protection status can act as a label indicating the quality of the resource, which is important for developing a competitive tourism product (Svalastog, 1998). This is true, not least for sub-polar and polar areas, which exhibit high numbers of national parks and protected areas because of (at least formerly) limited land-use competition and high wilderness values (Marsh, 2000; Hall and Boyd, 2005; Lemelin and Johnston, 2008; Müller, 2013a). Hence, in countries like Sweden, governments and authorities have identified national parks as resources for sustainable development and growth in peripheral regions (Swedish Environmental Protection Agency, 2004). This development is in line with recent changes in protection policies across Europe, marking a shift towards management models integrating the protection and use of parks and protected areas (Mose and Weixelbauer, 2007).

The aim to boost rural economic development based on national parks in Sweden has not been without criticism. Lundmark (2005) and Lundmark *et al.* (2010) argue, for example, that national parks in northern Sweden did not have the anticipated positive impacts on restructuring, employment and regional development. However, a positive impact of tourism on goods and service supply was recorded (Löffler, 2007). Even the relative importance of tourism for the labour market has increased since the 1960s, mainly because of changes in other sectors (Müller and Ulrich, 2007). However, since 2005, mining and other resource-extracting industries have increased in relative importance, while, for the first time, tourism has experienced a decline (Müller, 2013b). Opposition to the establishment of national parks in the Nordic countries exists not only because of doubtful economic impacts, but also because of the struggle for land-use rights and power (Puhakka *et al.*, 2009; Haukeland *et al.*, 2011; Kaltenborn *et al.*, 2011; Müller, 2013a).

With respect to tourism's role in regional development, Schmallegger and Carson (2010)

DOI: 10.1079/9781800621480.0006

warn about superficially successful tourism solutions. They argue that tourism-based economies can easily end up suffering from the same shortcomings as other mono-structured and export-oriented staple economies. Saarinen (2003, 2007) also points at the necessity of regarding tourism as one of many industries necessary to achieve a beneficial development. Therefore, linkages between tourism and other industries are more likely to achieve sustainable development than purely tourism-oriented approaches, which tend to result in vulnerable mono-structures (Kauppila *et al.*, 2009).

Against this background, the purpose of this chapter is to assess the extent to which tourism labour markets are interlinked with other sectoral labour markets in national park locations in northern Sweden and the extent to which mobility occurs between different sectors of the economy. The study area is the municipality of Jokkmokk in northern Sweden, with the Laponia World Heritage Site and numerous minor protected areas within its municipal limits.

National Parks and Tourism Development

It has not been established why national parks are often seen as resources for development. Certainly, national park status implies visibility on an international market and thus a greater likelihood of becoming an attraction (Wall *et al.*, 2007). Nature-based tourism in protected areas is often considered an economic driver, not least for the gateway communities into the area (Moisey, 2002). Roughly, positive impacts are closely related to visitor spending, resulting in indirect impacts on local tourism business, indirect impacts through backward linkages to other local businesses and induced impacts originating from the increased spending of employees in the area (Moisey, 2002; Job *et al.*, 2009). A particular problem in this context is leakage from the wider park area, including the gateway communities to other regions.

However, few studies actually prove the positive economic impacts of national parks in Europe. A common denominator for these studies is that they usually originate from North America, Australia or the developing world (cf. Moisey, 2002). In these regions, parks are often located far from major demand markets, require entry fees and offer a variety of services. The situation in European parks frequently differs (Mayer *et al.*, 2009) and even here national parks can frequently be found in relatively remote regions, although entry fees seldom apply and consumption opportunities for services and souvenirs might be more limited. This is particularly true in the Nordic countries, where the right of public access also comprises national parks and nature reserves (Kaltenborn *et al.*, 2001). This implies not only problems regarding management, but also worse preconditions for developing commercial services apart from accommodation and restaurants around the parks, since low-cost alternatives are available. Lundmark *et al.* (2010) notes that positive economic impacts can, thus far, not be recorded in the immediate neighbourhood of the parks. Fredman and Yuan (2011) register that economic impacts often occur outside the immediate surroundings of the parks and gateway communities and thus call for a wider spatial reach in studies of economic impacts in order to fully understand the importance of national parks and other protected areas. Hence, in order to leverage positive economic impacts, Mayer *et al.* (2009) identify two alternate routes: first to increase the number of visitors and second to lure them into spending more money on services, which often implies a qualitative upgrading of the services available. These avenues seem difficult to follow, however, since the peripheral location of national parks *per se* often constrains the number of tourists. Moreover, tourism entrepreneurship in these peripheral areas is also sometimes already far greater than is feasible, simply owing to a lack of alternative livelihoods (Lundmark and Müller, 2010). The Nordic countries are high-cost destinations (Svalastog, 1992) and thus higher prices may be counterproductive, since visiting tourists obviously prefer to spend relatively little money on their vacations (Mayer *et al.*, 2009; Fredman and Yuan, 2011). This further implies that new and probably smaller market segments have to be attracted.

However, it can be argued that there is indeed a third way to leverage the positive impacts of tourism: by diminishing leakage and creating other kinds of positive development beyond monetary measurements. This approach is likely to achieve limited economic yield, but, in contrast, it counteracts the fears expressed by Mayer *et al.* (2009) that in most cases increased tourism also means increasing environmental problems; moreover, the risk that seasonality will become even more prevalent is offset (Lundmark, 2006). Diminished leakage implies more tourists spending time within the area so that tourist spend spills over into other sectors of the local economy. This is in line with the ideas of Saarinen (2003, 2007), Kaupplia *et al.* (2009) and Schmallegger and Carson (2010). From their point of view, sustainable development is more about a diversification of local economies and a sound balance between tourism and other sectors, enabling communities to persist during crises, rather than simply about a booming tourism industry.

Mobility and Development

In economic geography, the occurrence of development and the ability to innovate are often explained by agglomerations and clusters, implying intensive interaction and co-operation between stakeholders (Malmberg, 1996; Porter, 1998). However, it is debatable as to what extent related or unrelated variety within the regional economy is more advantageous for the region's performance (Beaudry and Schiffauerova, 2009). One problem is how to empirically establish and delimit agglomerations and transactions between different companies occurring there. In this context, the mobility of personnel has been identified as something that can be captured and regarded as of particular importance, since people are seen as the major carriers of knowledge and innovation (Almeida and Kogut, 1999; Maskell and Malmberg, 1999; Boschma *et al.*, 2009; Eriksson and Lindgren, 2009). The literature suggests that new personnel implies a transfer of knowledge between firms, deepening the relationship between them. Not least, tacit knowledge is transferred mainly by labour mobility. For Swedish industries, Eriksson (2011) showed that spillovers involving dissimilar knowledge turned out to be useful for economic performance when occurring within a region, while similar knowledge spillovers turned out to be advantageous when transferred over long distances. Frenken *et al.* (2007) argue that knowledge spillover mainly results from related variety within a region and conclude that variety is positive as long as it occurs within the same industry. None of the above studies applies this logic to tourism, which usually remains unclassified within national statistical systems. Still, if it is assumed that the tourism industry functions like other industries, the above results imply that a spillover from tourism to other sectors, or *vice versa*, within a region should cause positive impacts on the economic performance of the region as long as these other industries are somewhat similar.

Thus far, tourism geographies lack these large-scale studies of labour market mobilities; additionally, the small-scale and scattered characteristics of the tourism industry and the specifics of tourism production make it far from self-evident that research findings from economic geography can be applied to tourism as well. However, it has been established, for example, that incomers are more often represented among tourism entrepreneurs (Paniagua, 2002). Moreover, Müller (2006) showed that tourism employment lured young people into remote areas of Sweden. After a couple of years many incomers had changed employment, working in sectors other than the tourism industry, showing at least some evidence of tourism's non-monetary impacts on community. Even Brouder (2012) reports on positive impacts of tourism on communities. In his study of Jokkmokk, however, the focus is on the intangibles of development, highlighting aspects such as local atmosphere and creativity, at least partly created by the development of tourism.

To achieve a development that deserves the label 'sustainable', the aspects discussed above should be addressed. In the case of remote national parks and protected areas, the environmental question in most cases appears to be mainly related to transportation and the number of tourists visiting the area, with manageable on-site impacts because of the overall low number of tourists.

Methodology

This study offers a case study quantitative approach to the assessment of labour mobility within tourism, used successfully in previous studies (Müller, 2006; Müller and Ulrich, 2007). Although a quantitative approach does not necessarily reveal motivations for and consequences of labour mobility, it has the advantage that it shows the overall patterns of change. Hence, a quantitative approach cannot answer all questions related to the impact of national parks on local communities. Nevertheless, it certainly illustrates the actual overall change and the circumstances in which this occurs.

The study uses data derived from the longitudinal geo-referenced micro-database ASTRID, located at the Department of Geography and Economic History, Umeå University. The database has been compiled by Statistics Sweden especially to allow research on labour market and population and features raw data that are used by Statistics Sweden to develop all domestic statistical information. This means that longitudinal information on a great number of demographic, social and economic issues is available for all individuals in Sweden between 1960 and 2007.

For this study, information on all individuals residing in Jokkmokk was retrieved for the time period 1990–2007. The focus was on information regarding place of residence and employment. For the place of residence, information on the municipality of residence was chosen to discern migrants into and from the municipality of Jokkmokk during the study period. The employment variables, containing annual information on the Swedish Industrial Classification (SIC) code of the workplace, enabled the detection of labour mobility. Finally, a distinct code signifying individual workplaces was retrieved in order to be able to associate single individuals with firms. Age was used to delimit the population. The problem of seasonal population and workforce was handled in respect of the Swedish situation, whereby income tax is paid in the municipality of permanent residence. In accordance, relatively little economic impact – daily consumption – is thus retained within the municipality of work. The same applies to commuters who are not numerous, owing to the relatively long distances between population centres. Hence, both categories were excluded, which is a certain shortcoming of the study.

A further challenge was the compilation of the tourism industry. In Sweden, like elsewhere, tourism is not classified as a single industry. Tourism Satellite Accounts (TSA) are therefore sometimes used to assess the share of tourism in the value creation within classified industries (Meis, 2001). TSA assessments of tourism shares in various industries have also been used for estimating tourism-induced employment (Müller, 2006). Of course, these approaches can be criticized for being inaccurate, since estimates of the relative importance of tourism are usually based on national averages and may thus be a poor reflection of the situation in certain places. Therefore, another approach was chosen here. All employment categories, based on the SIC five-digit code that clearly refers to employment characterized as being the core of the tourism industry, were selected and aggregated to represent the local tourism industry (Table 6.1). This approach has been used previously by Lundmark (2005, 2006). It should be noted, however, that not all codes could be found for the case of Jokkmokk. All other industries were summarized for the first digit of the SIC code.

The study further relies on a descriptive approach. To this end, public statistics on commercial overnight stays and changes in the industry and inter-sector mobility on the local labour market were mapped. More analytical statistics could certainly have been applied; however, the total scope of the data and the case study approach do not allow for comprehensive generalization departing from the data and hence more sophisticated statistics appeared to be obsolete.

Geographical Background

This study focuses on the municipality of Jokkmokk in northern Sweden. Located in the Arctic Circle, the municipality has recently suffered from economic decline and other problems related to its peripheral location. With roughly 5200 inhabitants and an area of 19,474 km^2, the municipality is among the most sparsely populated areas even in northern Sweden. Jokkmokk is located 170 km from the county capital of Luleå, where the closest major airport is located. It is possible to reach the centre of the municipality through combined train and bus transportation; however, using a car is the easiest way to reach Jokkmokk and its national parks. The municipality stretches from the Scandinavian mountain range in the west to the

Table 6.1. Tourism SIC codes according to 2002 standard.

Code	Key
55101	Hotels with restaurant except conference centres
55102	Conference centres with lodging
55103	Hotels and motels without restaurant
55210	Youth hostels etc.
55220	Camping sites including caravan sites
55230	Other short-stay lodging facilities
55300	Restaurants
55400	Bars
55521	Catering to the transport sector
61200	Inland water transports
62100	Scheduled air transports
62200	Non-scheduled air transports
63301	Activities of tour operators
63302	Activities of travel agencies
63303	Tourist assistance
92320	Operation of arts facilities
92330	Fair and amusement park activities
92340	Other entertainment activities
92520	Museum activities and preservation of historical sites and buildings
92530	Botanical and zoological gardens and nature reserve activities
92611	Operation of skiing facilities
92612	Operation of golf courses
92729	Various other recreational activities
52485	Retail sale of sports and leisure goods

boreal forest landscape of eastern Scandinavia. The municipality has long been a centre for the Sami, the indigenous population of northern Europe and their reindeer herding activities. However, forestry and hydropower have meant important changes to local industry. Moreover, in 1909 the municipality became the site of two of Europe's first National Parks, Sarek and Stora Sjöfallet. Later, the Muddus (1942) and Padjelanta (1962) National Parks were added. These IUCN II-category parks and a number of nature reserves, together with the presence of the indigenous Sami reindeer herding culture, were the motivation for conferring the status of UNESCO World Heritage Site on the Laponia area in 1996. This was expected to boost the tourism industry. Presently, 35% of the municipality is protected nature. In the Swedish case, this implies a transfer of planning authority for the protected areas from the municipality to the state, limiting local ability to engage in development projects other than those related to nature protection (Lundmark and Stjernström, 2009).

The parks are main attractions for tourism in Jokkmokk. Indeed, the fact that Carl Linnæus ended his Iter Lapponicum in Kvikkjokk, a mountain village in today's municipality, is frequently mentioned as the embryo of modern mountain tourism in Sweden. Today, several hiking trails pass the parks with huts offering accommodation to tourists. However, Sarek National Park lacks more or less all touristic infrastructure. It should also be noted that in Sweden, national parks are not managed like they are, for example, in North America; there are no entrance fees and no employees catering to the needs of tourists. As compensation for the hydropower development in the municipality, Ájtte (the Swedish Mountain and Sami Museum) was built in Jokkmokk in the early 1990s and forms an additional attraction, not least for car-based tourists. Moreover, Sami tourism is an important attraction (Müller and Kuoljok Huuva, 2009). The Jokkmokk Winter Festival, held in early February and with a more than 400-year-old history, is a major tourism event, attracting more than 30,000 visitors during just three days (Pettersson, 2003).

As in many other parts of northern Sweden, tourism figures have in fact declined during the past 20 years: commercial accommodation figures dropped from 92,000 in 1990 to 75,000 in 2006, with foreign tourists representing 19% and 17.5%, respectively. Moreover, during the same period tourism changed dramatically; previously mostly a summer destination, the area is increasingly profiling itself around winter products, implying two seasons and improved preconditions for tourism businesses (Müller, 2011). A development towards increasing tourism is also being planned for within the municipality. Hence, in 2011, a new destination management organization (DMO), Destination Jokkmokk, was established to improve and co-ordinate tourism development and increase the number of tourists to the municipality.

Results

The results section scrutinizes three aspects of tourism development in Jokkmokk. First, the statistics on commercial overnight stays are analysed in order to discern publicly anticipated impacts of the World Heritage designation and the national parks. Next, the local tourism industry is assessed with regard to the number of firms and employees. Finally, data on mobility within the local tourism labour market are presented in order to find other positive impacts related to skill transfers and in-migration.

Changes in tourism demand

As already mentioned, commercial tourism figures decreased during the research period. However, this development does not mirror the development in the county as a whole. In Norrbotten county, commercial overnight stays increased by almost 40% from 1990 to 2006 (Fig. 6.1). Not least, the positive tourism development in Kiruna, related to the establishment of the *Icehotel* in the early 1990s and the tourism development related to car testing under winter conditions in Arvidsjaur, explain the positive development in the county. These developments also mark the establishment of a winter season with international appeal in an area that had previously attracted mainly domestic summer tourists (Müller, 2011). Furthermore, both destinations have airline connections to Stockholm (and during the winter season also abroad), providing a competitive advantage.

During the same time period in Jokkmokk municipality, the number of commercial overnight stays dropped by roughly 20%. It was only in 1991 and 1992 that Jokkmokk showed a more beneficial trend than the county did. The number of overnight stays measured in 1990 could only roughly be equalled in 1992 and 1995. Although the development in Jokkmokk mirrors annual variations in the county figures, particularly a drop between 1992 and 1993, a period of economic crisis in Sweden and strong declines between 1995 and 1997, as well as after 2003, created less favourable conditions for tourism development in the municipality. The Jokkmokk Winter Festival, an annual major weekend event in early February, generates numerous overnight stays in Jokkmokk. However, the overwhelming majority of these stays are in private homes and are thus not included in the official statistics (Pettersson, 2003). Still, the very positive development of the event has not entailed increased visitation during other seasons. Even the World Heritage Site designation in 1996 had no measurable impact on the number of visitors staying overnight in the area. In fact, figures after 1996 reached the same level again only once, in 2003.

Obviously, fluctuations in overnight figures are similar for the county and the municipality. It can thus be noted that the establishment of the Laponia World Heritage area did not make a significant difference to overnight stays in the

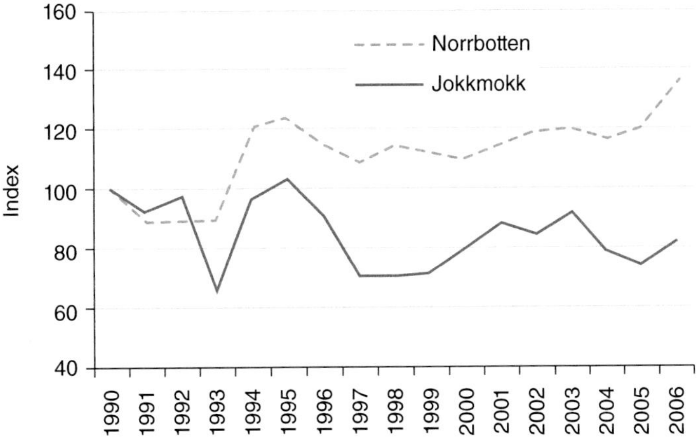

Fig. 6.1. Indexed development of commercial overnight stays in Norrbotten county and Jokkmokk municipality, 1990–2006. (From Statistics Sweden, 2007; Regionfakta, 2012)

municipality. This might be explained by the fact that Laponia is mainly a summer destination, as no major winter tourism products could be developed. This could be because of the harsh winter conditions in the partly alpine area, but also because of the restrictions that apply in the national park areas. Accordingly, a ban on driving snowmobiles in protected areas, an otherwise popular leisure activity in winter in Sweden, may explain the development. Thus, the local tourism supply and also rather limited accessibility failed to meet the new requirements of increasing international demand.

Changes in the tourism industry

An increasing engagement within the tourism industry can be caused by push as well as pull factors. A general decline in job opportunities owing to restructuring may force labour into economic sectors that persist because of their place-based characteristics. Tourism, particularly nature-based tourism, is one of the industries that is based on scenic and environmental resources, which are relatively constant. Still, demand varies because of changing demand patterns caused by fashion and economic fluctuations. However, tourism also has the potential to pull people into the industry. Economic and lifestyle considerations are reasons to become employed within the industry (Carson *et al.*, 2018).

Indeed, scrutinizing the number of tourism enterprises in Jokkmokk reveals that the number of tourism firms increased during the study period (Fig. 6.2). Particularly after 1998, two years after the establishment of the World Heritage area, the establishment of new tourism companies took off: the total number increased from 22 to 40 within a nine-year period.

During the same period, the average workplace size decreased from 8.4 to 4.9 workers. This is in sharp contrast to the period prior to 1998, when tourism companies were fewer but larger. For example, in 1993, a total of 20 workplaces had an average of 13.8 workers. There has therefore been an increasing fragmentation of the local tourism industry into smaller enterprises. Considering the growing requirements of competences within the tourism industry, it is obvious that the Jokkmokk companies have become smaller and thus less likely to be competitive in an international market. However, the statistics on companies showing an increase in activity disguise the fact that there has been a decline in the number of tourism workers. Starting from the low number of 184 in 1990, the numbers peaked in 1995, a year prior to the World Heritage Site designation, when 291 people worked within the tourism industry (Fig. 6.3).

Mobility in the local labour market

This corresponds to the development of overnight stays in the municipality. After 1995, the tourism workforce decreased rapidly and recovered

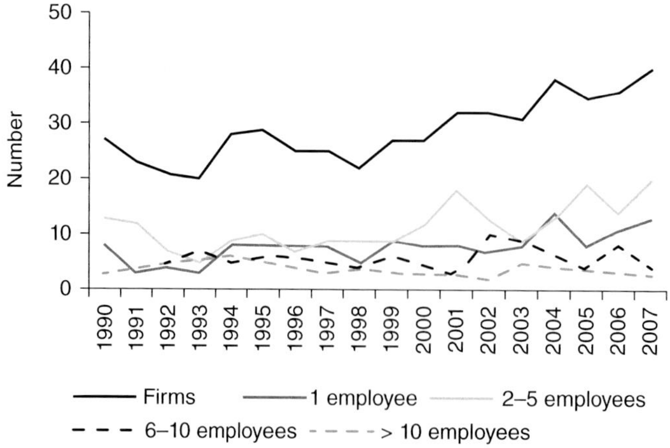

Fig. 6.2. Number of tourism firms in Jokkmokk based on size, 1990–2007. (From ASTRID database)

to a level of roughly 200 people. This development was initially simultaneous with that of the role of tourism in the local labour market. In 1990, the tourism workforce was 5.3% of the total workforce in Jokkmokk and in 1995 it represented 8.7%. After 1995, the development of the size of the workforce and of its relative share of the local labour market diverged somewhat. The shrinking of the tourism workforce is mirrored only partly in the decline of its relative importance. Hence, between 2001 and 2004, the relative share of tourism labour again exceeded 7%, consecutively dropping to a level of about 6.5% in 2007. This shows that the relative importance of tourism for the local labour market increased, although the number of people employed within tourism continued to shrink. This means that other sectors of the local labour market declined more rapidly than tourism. Still, there is no indication that the establishment of the World Heritage area generated a greater number of job opportunities. However, it apparently stimulated a willingness to start new enterprises within tourism.

A final argument regarding the potential importance of tourism has been related to its impact on other sectors in the form of knowledge spillover. First, to illustrate the conditions in Jokkmokk, overall mobility patterns for the tourism sector are demonstrated. Then, the analysis focuses on the other industry sectors involved.

It can be noted that mobility in the tourism labour market was more frequent during the early 1990s, prior to 1996 (Fig. 6.4). In 1991, a total of 129 people, previously in other occupations, were employed within tourism; 73 of them were registered in another municipality in 1990 representing an external inflow into Jokkmokk's tourism industry, while 56 had moved from another sector within the municipality. The combined inflow corresponds to 70% of tourism labour employed in 1990. Thus, the sector expanded significantly from 1990 to 1991, as was demonstrated earlier (cf. Fig. 6.3). Until 1995, inflow remained over 40%, causing a great growth in employment within tourism. Afterwards, levels stabilized at around 35%, with a majority entering the tourism sector in Jokkmokk from outside the municipality.

There was an outflow from the tourism industry that superseded the inflow between 1996 and 1998 as well as after 2003. The focus here is on outflows to other local industries. Except for 1991 and 1992, the local outflow was between 10% and 20% annually. Levels were lower during the last five years of the covered research period, when between 20 and 30 people left the tourism industry for other occupations within Jokkmokk municipality. Hence, particularly after the establishment of the World Heritage area, the potential impact of tourism on other sectors of the local labour market appears to be limited. The general impression is that employment within tourism is highly volatile.

As argued earlier, however, one important question is not necessarily how many people

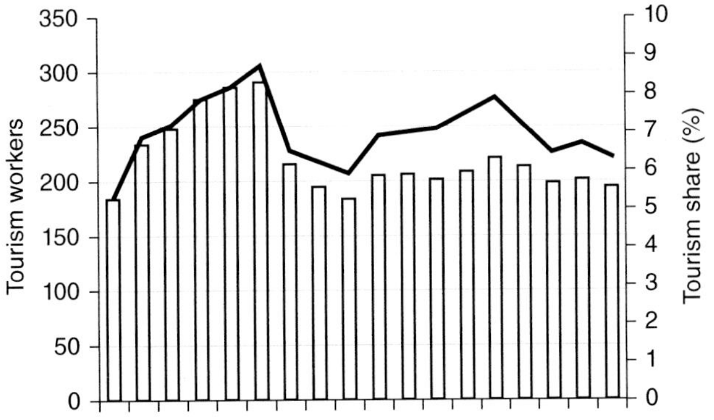

Fig. 6.3. Number of tourism workers (columns) and share of tourism workers in local workforce (line), 1990–2007. (From ASTRID database)

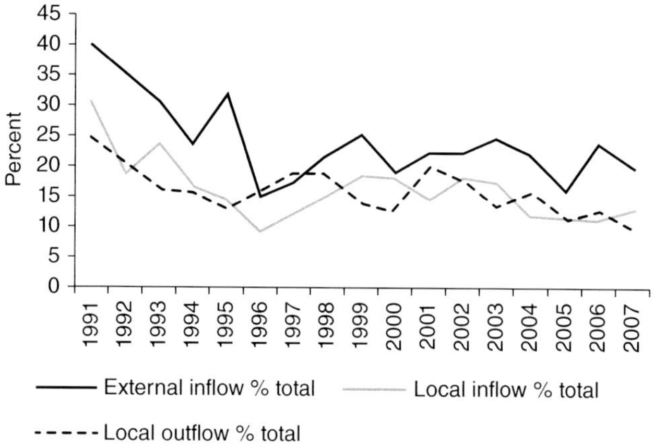

Fig. 6.4. Annual inflows and outflows as a percentage of the total annual workforce in Jokkmokk, 1991–2007. (From ASTRID database)

enter and leave the tourism industry, but rather what kind of experiences they import and export from the industry to other sectors of the economy. Regarding recruitment, the largest group of people entering the tourism industry derives from unemployment or school (Fig. 6.5). This is particularly true for people moving into the municipality. Of a total of 900 individuals, 362 were inactive on the labour market prior to moving to Jokkmokk, while for domestic flows, 115 of 600 people had had no other occupation.

The second largest group is recruited from non-economically productive sectors, namely healthcare and education. A minority is recruited from related sectors such as non-tourism hospitality, for example school catering and retail trade. Consequently, it appears that previous knowledge and work experience are not necessarily essential for employment in the tourism industry in Jokkmokk.

The patterns of local outflow largely mirror the situation mapped for inflows. Even here, non-economically productive service industries are the destinations for the workforce previously employed in tourism. The health care sector and education together receive roughly a third of all people leaving the tourism industry. The third largest group enters unemployment or retirement. It can therefore be noted that a majority of those leaving the local tourism industry for other occupations in the municipality in fact end up in public employment. These industries are only loosely related to tourism. For the Swedish situation, this means that people leaving tourism enter sectors characterized by low income but relatively stable demand. As a result, it appears likely that low-income jobs in a volatile tourism industry are exchanged for low-income jobs in sectors offering stable demand and non-seasonal work.

Discussion

The analysis of quantitative data for this case study creates a puzzle clearly indicating that the tourism industry in Jokkmokk at large is contested. In contrast to public expectations, the national parks and the establishment of a World Heritage area in 1996 have been insufficient to alter a negative development trend for tourism in the municipality. Instead, numbers of employees and guest nights continue to decline; only the number of enterprises classified as touristic is increasing. However, these are mainly micro-firms and it appears likely that they are rather an expression of overall decline and a situation in which clutching at straws, in this case within tourism, is considered a way to sustain a livelihood and residence in a marginal municipality.

Considering the overall positive tourism development of Norrbotten county, it is reasonable to ask why Jokkmokk departs from the general trend despite having access to premier environments designated as nature reserves, national

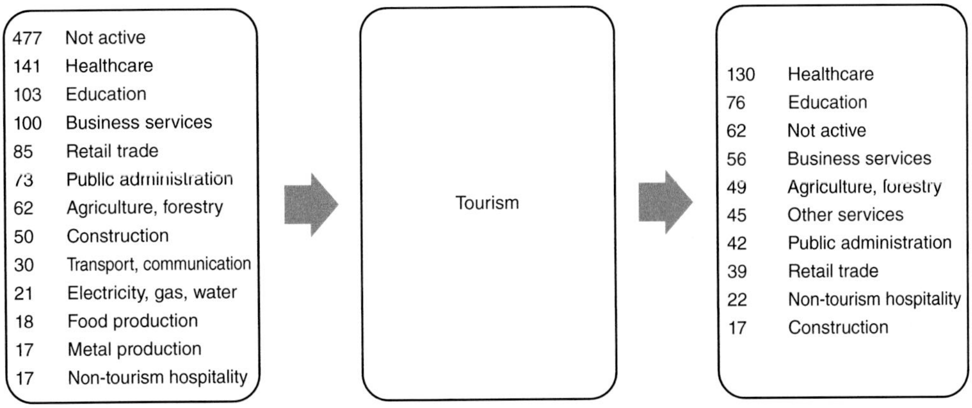

Fig. 6.5. Volume (absolute numbers) of inflows and local outflows to and from the tourism labour market in Jokkmokk, 1991–2007. (From ASTRID database)

parks and a World Heritage Site. Accessibility is obviously an important issue, as is the establishment of a winter season (Müller, 2011). Here, Jokkmokk is disadvantaged owing to its lack of nearby airports.

Mayer *et al.* (2009) argue that national parks can entail positive economic impacts in two different ways. Obviously, the first – aiming to increase tourism numbers – failed, and, instead, a significant decline has to be reported. The second impact is related to visitor spending. Neither this nor the degree of leakage could be mapped in this case study. However, the obvious shrinkage of the tourism workforce during the study period indicates that no major increase in spending seems to have occurred. As argued previously, a positive impact can also be related to a knowledge spillover into other related sectors. The study indicated that this does not seem to be the case, however, since it is primarily the public sector and companies that receive the workforce leaving tourism, which is characterized by low income levels. Since these still appear to be attractive, this reveals the poor situation for tourism that cannot even compete with low-income employment within the public sector.

Still, Brouder (2012) argues that there are positive tourism impacts to be seen regarding non-tangible aspects of community life in Jokkmokk. This may be the case, since the situation in the tourism sector is also certainly mirrored in other local industries. Hence, tourism may be seen as a last resort where employment can be created, which is also reflected in the increasing number of tourism firms. Moreover, the current study indicates the importance of tourism as an engine for attracting in-migration to the municipality. This has also been demonstrated previously in a similar case in northern Sweden (Müller, 2006). However, the development does not seem to have the power to alter otherwise negative population trends, and recently detected mineral resources may turn out to be an alternative development path for the municipality, as already recognized elsewhere in northern Sweden (Müller, 2011, 2013b). The perception of a northern pleasure periphery based on pristine nature may be truncated backwards to the harsh reality of a mineral resource hinterland.

Conclusions

This case study has shown that the national parks in Jokkmokk do not have any obvious positive impact on local labour markets. Rather, despite the presence of national parks, tourism in the area is developing more weakly than in adjacent municipalities with fewer protected areas and areas with lower protection status. This study therefore contests the idea that national parks are beneficial to regional development (Moisey, 2002). So far, at least, it seems to be doubtful that Hammer's (2007) notion of the new role of national parks beyond protection actually has any greater degree of practical implications contributing to sustaining peripheral communities. The least that can be said is that other factors, such as accessibility and a viable product mix, seem to be far more important for turning a nature area into a resource, than the availability of protected nature. This is in line with Svalastog's (1998) argument also highlighting the role of infrastructures and suprastructures. A policy shift indicating new possibilities for using protected nature alone is insufficient to counteract locational disadvantages.

However, the study indicated that national parks may have a role to play in that they at least offer an opportunity to develop new businesses. Moreover, tourism is seemingly an important labour market when it comes to attracting in-migration and functions as an escalator, bringing new people into the municipality. Still, many in-migrants seem to move on quickly to other sectors of the local economy, which supports earlier findings (Müller, 2006). However, this development seems to be limited to public sector jobs such as healthcare. The tourism labour market appears to remain rather isolated in the local economy and the high degree of in-migrated labour further constrains potential linkages between different sectors of the local economy. Against this background, it can be noted that tourism development does not appear to become an integrated part of local business life and its potential to achieve a more sustainable development as required by Kauppila *et al.* (2009), thus far, remains rudimentary.

References

Almeida, P. and Kogut, B. (1999) Localization of knowledge and the mobility of engineers in regional networks. *Management Science* 45(7), 905–917.

Baum, T. (1999) The decline of the traditional North Atlantic fisheries and tourism's response: the cases of Iceland and Newfoundland. *Current Issues in Tourism* 2(1), 68–81.

Beaudry, C. and Schiffauerova, A. (2009) Who's right, Marshall or Jacobs? The localization versus urbanization debate. *Research Policy* 38(2), 318–337.

Boschma, R., Eriksson, R. and Lindgren, U. (2009) How does labour mobility affect the performance of plants? The importance of relatedness and geographical proximity. *Journal of Economic Geography* 9(2), 169–190.

Brouder, P. (2012) Creative outposts: tourism's place in rural innovation. *Tourism, Planning & Development* 9(4), 383–396.

Buckley, R. (2004) The effects of World Heritage listing on tourism to Australian National Parks. *Journal of Sustainable Tourism* 12(1), 70–84.

Carson, D.A., Carson, D.B. and Eimermann, M. (2018) International winter tourism entrepreneurs in northern Sweden: understanding migration, lifestyle, and business motivations. *Scandinavian Journal of Hospitality and Tourism* 18(2), 183–198.

Eriksson, R.H. (2011) Localized spillovers and knowledge flows: How does proximity influence the performance of plants? *Economic Geography* 87(2), 127–152.

Eriksson, R.H. and Lindgren, U. (2009) Localized mobility clusters: impacts of labour market externalities on firm performance. *Journal of Economic Geography* 9(1), 33–53.

Fleischer, A. and Felsenstein, D. (2000) Support for rural tourism: Does it make a difference? *Annals of Tourism Research* 27(4), 1007–1024.

Fredman, P. and Yuan, M. (2011) Primary economic impacts at three spatial levels: the case of Fulufjället national park, Sweden. *Scandinavian Journal of Hospitality and Tourism* 11(Suppl.), 74–86.

Fredman, P., Hörnsten Friberg, L. and Emmelin, L. (2007) Increased visitation from national park designation. *Current Issues in Tourism* 10(1), 87–95.

Frenken, K., Van Oort, F. and Verburg, T. (2007) Related variety, unrelated variety and regional economic growth. *Regional Studies* 41(5), 685–697.

Hall, C.M. (2007) North-south perspectives on tourism, regional development and peripheral areas. In: Müller, D.K. and Jansson, B. (eds) *Tourism in Peripheries: Perspectives from the Far North and South*. CAB International, Wallingford, UK, pp. 19–37.

Hall, C.M. and Boyd, S. (2005) Nature-based tourism in peripheral areas. Introduction. In: Hall, C.M. and Boyd, S. (eds) *Nature-based Tourism in Peripheral Areas: Development or Disaster?* Channel View, Clevedon, UK, pp. 3–17.

Hammer, T. (2007) Protected areas and regional development: conflicts and opportunities. In: Mose, I. (ed.) *Protected Areas and Regional Development in Europe: Towards a New Model for the 21st Century*. Ashgate, Aldershot, UK, pp. 21–36.
Haukeland, J.V., Daugstad, K. and Vistad, O.I. (2011) Harmony or conflict? A focus group study on traditional use and tourism development in and around Rondane and Jotunheimen national parks in Norway. *Scandinavian Journal of Hospitality and Tourism* 11(Suppl.), 13–37.
Jenkins, J.M., Hall, C.M. and Troughton, M. (1998) The restructuring of rural economies: rural tourism and recreation as government response. In: Butler, R., Hall, C.M. and Jenkins, J. (eds) *Tourism and Recreation in Rural Areas*. Wiley, Chichester, UK, pp. 43–68.
Job, H., Woltering, M. and Harrer, B. (2009) *Regionalökonomische Effekte des Tourismus in deutschen Nationalparken*. Bundesamt für Naturschutz, Bonn, Germany.
Jóhannesson, G.T. and Huijbens, E.H. (2010) Tourism in times of crisis: exploring the discourse of tourism development in Iceland. *Current Issues in Tourism* 13(5), 419–434.
Kaltenborn, B.P., Haaland, H. and Sandell, K. (2001) The public right of access: some challenges to sustainable tourism development in Scandinavia. *Journal of Sustainable Tourism* 9(5), 417–433.
Kaltenborn, B.P., Qvenlid, M. and Nellemann, C. (2011) Local governance of national parks: the perception of tourism operators in Dovre-Sunndalsfjella national park, Norway. *Norsk Geografisk Tidsskrift* 65(2), 83–92.
Kauppila, P., Saarinen, J. and Leinonen, R. (2009) Sustainable tourism planning and regional development in peripheries: a Nordic view. *Scandinavian Journal of Hospitality and Tourism* 9(4), 424–435.
Lemelin, R.H. and Johnston, M. (2008) Northern protected areas and parks. In: Dearden, P. and Rollins, R. (eds) *Parks and Protected Areas in Canada: Planning and Management* (3rd edn). Oxford University Press, New York, pp. 294–313.
Löffler, G. (2007) The impact of tourism on the local supply structure of goods and services in peripheral areas: the example of northern Sweden. In: Müller, D.K. and Jansson, B. (eds) *Tourism in Peripheries: Perspectives from the Far North and South*. CAB International, Wallingford, UK, pp. 69–84.
Lundmark, L. (2005) Economic restructuring into tourism in the Swedish mountain range. *Scandinavian Journal of Hospitality and Tourism* 5(1), 23–45.
Lundmark, L. (2006) Mobility, migration and seasonal tourism employment: evidence from Swedish mountain municipalities. *Scandinavian Journal of Hospitality and Tourism* 6(3), 197–213.
Lundmark, L. and Müller, D.K. (2010) The supply of nature-based tourism activities in Sweden. *Tourism* 58(4), 379–393.
Lundmark, L. and Stjernström, O. (2009) Environmental protection: an instrument for regional development? National ambitions versus local realities in the case of tourism. *Scandinavian Journal of Hospitality and Tourism* 9(4), 387–405.
Lundmark, L.J.T., Fredman, P. and Sandell, K. (2010) National parks and protected areas and the role for employment in tourism and forest sectors: a Swedish case. *Ecology and Society* 15(1), article 19.
Malmberg, A. (1996) Industrial geography: agglomeration and local milieu. *Progress in Human Geography* 20(3), 392–403.
Marsh, J. (2000) Tourism and national parks in polar regions. In: Butler, R.W. and Boyd, S.W. (eds) *Tourism and National Parks: Issues and Implications*. Wiley, Chichester, UK, pp. 125–136.
Maskell, P. and Malmberg, A. (1999) Localised learning and industrial competitiveness. *Cambridge Journal of Economics* 23(2), 167–185.
Mayer, M., Müller, M., Woltering, M., Arnegger, J. and Job, H. (2009) The economic impact of tourism in six German national parks. *Landscape and Urban Planning* 97(2), 73–82.
Meis, S. (2001) Towards comparative studies in Tourism Satellite Accounts. In: Lennon, J.J. (ed.) *Tourism Statistics: International Perspectives and Current Issues*. Continuum, London, pp. 14–23.
Moisey, R.N. (2002) The economics of tourism in national parks and protected areas. In: Eagles, P.F.J. and McCool, S.F. (eds) *Tourism in National Parks and Protected Areas: Planning and Management*. CAB International, Wallingford, UK, pp. 235–253.
Mose, I., and Weixelbaumer, N. (2007) A new paradigm for protected areas in Europe? In: Mose, I. (ed.) *Protected Areas and Regional Development in Europe: Towards a New Model for the 21st Century*. Ashgate, Aldershot, UK, pp. 3–20.
Müller, D.K. (2006) Amenity migration and tourism development in the Tärna Mountains, Sweden. In: Moss, L.A.G. (ed.) *The Amenity Migrants: Seeking and Sustaining Mountains and Their Cultures*. CAB International, Wallingford, UK, pp. 245–258.

Müller, D.K. (2011) Tourism development in Europe´s 'last wilderness': an assessment of nature-based tourism in Swedish Lapland. In: Grenier, A.A. and Müller, D.K. (eds) *Polar Tourism: A Tool for Regional Development*. Presses de l'Université du Québec, Montreal, Canada, pp. 129–153.

Müller, D.K. (2013a) National parks for tourism development in sub-Arctic areas – curse or blessing? The case of a proposed national park in northern Sweden. In: Müller, D.K., Lundmark, L. and Lemelin, R.H. (eds) *New Issues in Polar Tourism: Communities, Environments, Politics*. Springer, Dordrecht, The Netherlands, pp. 189–204.

Müller, D.K. (2013b) Hibernating economic decline? Tourism and labor market change in Europe's northern periphery. In: Visser, G. and Ferreira, S. (eds) *Tourism and Crisis*. Routledge, London, pp. 113–128.

Müller, D.K. and Kuoljok Huuva, S. (2009) Limits to Sami tourism development: the case of Jokkmokk, Sweden. *Journal of Ecotourism* 8(2), 115–127.

Müller, D.K. and Ulrich, P. (2007) Tourism development and the rural labour market in Sweden, 1960–1999. In: Müller, D.K. and Jansson, B. (eds) *Tourism in Peripheries: Perspectives from the Far North and South*. CAB International, Wallingford, UK, pp. 85–105.

Paniagua, A. (2002) Urban-rural migration, tourism entrepreneurs and rural restructuring in Spain. *Tourism Geographies* 4(4), 349–371.

Pettersson, R. (2003) Indigenous cultural events: the development of a Sami winter festival in Northern Sweden. *Tourism* 51(3), 319–332.

Porter, M.E. (1998) *On Competition*. Harvard Business School, Boston, Massachusetts.

Puhakka, R., Sarkki, S., Cottrell, S.P. and Siikamaäki, S. (2009) Local discourse and international initiatives: sociocultural sustainability of tourism in Oulanka national park, Finland. *Journal of Sustainable Tourism* 17(5), 529–549.

Regionfakta (2012) Gästnätter mm månadsvis 2010 ff (hotell, stugbyar och vandrarhem): Norrbottens län. Available at: www.regionfakta.com/Norrbottens-lan/Snabbstatistik/Gastnatter-pa-hotell-stugbyar-och-vandrarhem (accessed 20 November 2018).

Saarinen, J. (2003) The regional economic of tourism in northern Finland: the socio-economic implications of recent tourism development and future possibilities for regional development. *Scandinavian Journal of Hospitality and Tourism* 3(2), 91–113.

Saarinen, J. (2007) Tourism in peripheries: the role of tourism in regional development in northern Finland. In: Müller, D.K. and Jansson, B. (eds) *Tourism in Peripheries: Perspectives from the Far North and South*. CAB International, Wallingford, UK, pp. 41–52.

Schmallegger, D. and Carson, D. (2010) Is tourism just another staple? A new perspective on tourism in remote regions. *Current Issues in Tourism* 13(3), 201–221.

Schmallegger, D., Harwood, S., Cerveny, L. and Müller, D. (2011) Tourist populations and local capital. In: Carson, D., Rasmussen, R.O., Ensign, P., Huskey, L. and Taylor, A. (eds) *Demography at the Edge: Remote Human Populations in Developed Nations*. Ashgate, Aldershot, UK, pp. 271–288.

Statistic Sweden (2007) *Tabell V:G H450 (1990-2006): Övernattande gäster efter hemland, 26 nationaliteter*. Örebro, Sweden.

Svalastog, S. (1992) Tourism in a changing Norway. *Norsk Geografisk Tidsskrift* 46(2), 109–126.

Svalastog, S. (1998) *The Nature of Resources for Tourism and Recreation. Working paper 68/1998*. Høgskolen i Lillehammer, Norway.

Swedish Environmental Protection Agency (2004) *Protect, preserve, present: a programme for better use and management of protected areas, 2005–2015 (Report 5483)*. Swedish Environmental Protection Agency, Stockholm.

Wall Reinius, S. and Fredman, P. (2007) Protected areas as attractions. *Annals of Tourism Research* 34(4), 839–854.

7 The Competitive Power in the Norwegian and Local Tourist Industry

Børge Dahle

What Are the Findings of the Research?

This is a summary of the research report 'Om den raskt avtakende konkurransekraft i norsk reiseliv' ('The rapidly declining competitive power in the Norwegian tourist industry') by Dr Sondre Svalastog (2010). His report is based on figures from the Central Bureau of Statistics.

The report includes the article 'Memorandum on research and education in the field of tourism', by Strømme Svendsen (2010), who was also the secretary of the group that produced the official report on *Research and Development Requirements in Village Tourism* (1988).

> Competitive power is not inherited – it is created. Knowledge determines what one is capable of getting out of what is accessible in the region, and relevant knowledge is decisive for the long-term administration and development of resources (goods). The production of experience is much more important for the tourism sector than would be indicated by the value share, seen in isolation. The sector of the tourist industry that is exposed to competition (based on resources and conditioned by leisure) is oriented to specific districts and linked to qualities in Norwegian nature and the cultural landscape (free goods).
>
> The Norwegian tourist industry today lacks the competence to understand, administer and develop the basic sphere of nature and the cultural landscape. One consequence of this is that the sector has no contacts or collaboration with the professional representatives that carry out the administration and that enable people to work with these attractions.
>
> The Norwegian tourist industry has concentrated its attention on the last link in the production chain (primarily: marketing, administration, financing, and leadership, including professional jargon), but this link has little influence on the quality and on the adaption to the market, and thus on the competitive power of the industry.

What were the consequences?

The Norwegian tourist industry made profits before the Second World War and during the 1960s. From 1970, the number of international tourists to Norway diminished considerably. In the 1960s, the Norwegian tourist industry made a significant profit on the size of the firms involved, improving the relationship between the scope of the production department and the administration.

> The share of international turnover halved between 1970 and 2009. Changes of this size can only occur when basic factors collapse absolutely, which is why quality and regional authenticity are so important. There has been little improvement in the Norwegian tourist industry and there is insufficient access to tourism business competence. Hence, the

 Tourism as a Resource-based Industry: Based on the work of Sondre Svalastog (eds A.L. Svalastog, D.K. Müller and I. Jenkins)
DOI: 10.1079/9781800621480.0007

development of competence must be linked to the relative competitive advantages of the country; that is to say, to the rich experiences offered by nature and its special culture.

What are the possibilities?

Every product and tourist attraction is a result of the quality of:

- raw materials (nature and culture);
- the workforce;
- capital.

> Some of Norway's raw materials are world-class (this applies especially to nature). Also, there is no shortage of capital in today's Norway. If this is accepted as a fact, one must look for explanations of this contemporary situation within the Norwegian tourist industry and workforce, which can be considered as having knowledge that makes tourism successful. In a high-cost country like Norway, production theory tells us that all production that is exposed to competition must be based on knowledge (Svalastog, 2010).
>
> Economic theory tells us that a rising level of salaries and costs must be met with a corresponding growth in the level of competence if one is not to lose competitive power. Products that are exposed to competition in a high-cost country therefore require a higher input of competence. (When I use the collective concept of 'competence', this includes not only knowledge and education, but also entities that are enigmatic: temperament, disposition and mentality.
>
> (Arnljot Strømme Svendsen, Professor Emeritus at the Norwegian Business College)

Sondre Svalastog's understanding and knowledge of the Norwegian tourist sector should be considered as an important foundation for developing the future of the local/regional tourist industry. Svalastog's conclusion can be summed up thus:

- In order to increase the competitive power in the Norwegian tourist industry, it is necessary to develop attractions and shared goods based on nature and culture.
- To be successful it is necessary to have competency in fully understanding, administering and developing natural resources and the cultural landscape.

What Can Be Done to Increase Competitive Power?

Svalastog's research report concludes that the limiting factor in the development of the tourist industry is competence linked to products and attractions offering experiences. This concurs with the considerations described in the article 'Nature and culture in leisure and industrial development: a local community that acts together' (Dahle, 1992).

It is important to note the point made by Strømme Svendsen: his understanding of competence includes not only knowledge but also 'things that are mysterious: temperament, disposition, and mentality'. It is surely beyond doubt that different cultures in different local government districts (*kommune* in Norwegian) can be seen as a result of history and of different presuppositions with regard to location, communication etc.

Experience and research indicate that the most decisive factor for a positive development of the tourist industry is the presence of an aggressive and optimistic development culture in the local government district. It is also important that there is an acceptance of the need for development (or further development) that is knowledge based; if no development culture exists, one must be created.

Creating a development culture depends on the active involvement of local business people and other actors within the community. These persons may be in the public or private sphere but what is of primary importance is that they are given freedom to act.

It will always be the case that initiatives taking place in a local government district will be initiated either externally or from within. These initiatives can form the basis of development and provide a stimulus to modernization. What counts is seizing an opportunity when it arises. This understanding accords with the findings of Telemark Research in its research project Successful Rural Local Government Districts.

It is vitally important to accept that a new creation must build on the development of competence; and raising the level of competence must be goal-oriented and action based. In other words, it underpins new project developments with the potential to reach an acceptable level of quality.

Against the background of Sondre Svalastog's research into 'the strategies that can increase the competitive power in the Norwegian tourist industry', we must identify and define the 'key' to establishing a competence in a group, which is responsible for developing a sustainable tourist industry in a local government district or within a real tourist project.

In this context, I refer to two studies that aim to identify strategies for raising the competence of those who are responsible for product development:

1. a study based on 'action-based development of competence' to strengthen the tourist industry *qua* industry; three different actions are described;
2. a report on how to strengthen the product quality in a project of national dimensions, 'The Pilgrim Path: an International Project in the Tourist Industry'.

Action-based development of competence

The 'action-based development of competence' can be organized as a formal study, as follow-up studies or as one method of approach in a developmental process. It can be linked to a process of development in the tourist industry, which includes, in more general terms, organizations and/or cultural projects. The development of competence takes place through a goal-oriented process that seeks to implement real projects. The focus is on the development and the quality of the product.

One important presupposition is that participants in the process intend to realize their 'idea' and that they are in a position to make this possible. The tangible project can be initiated through a creative analytical process in which a competent group arrives at an idea for a product. It can also be initiated by a competent person or persons.

Action-based development of competence is a strategy that ensures both implementation and quality. It has been implemented in ordinary studies at universities, in follow-up studies, in advanced training and in various developmental processes; and it has produced results. The visible results can be seen in established firms/actions, in developmental actions in organizations or in established competences in students.

Action-based development of competence is a process of developing the tourist industry in order to strengthen the competitive power in a hinterland local government district in Norway. A development process implemented in small local government districts or in one specific geographical area makes it possible to develop and implement real projects by making use of 'the advantages of small firms'; that is to say, through ideas, decisions and implementation. One such case study can be found as a developmental process in a local government district in the interior of Norway. The strategy of action-based development of competence was carried out in various projects over a long period of time in order to increase the competitive power of the tourist industry in this district.

This work was extremely challenging because the presuppositions in the local environment for real action were very limited. 'Temperament, disposition, and mentality' were involved to a large extent and there was a shortage of persons who wanted or were able to work on the development of attractions based on local nature and culture in a developmental process, which can be regarded as a 'journey'. There existed, however, a high 'basic competence', viewed in the context of knowledge of nature and culture in the local government district. It was important to develop the appreciation that an additional 'complementary' competence was needed to ensure new developments were created; it is the combination of various competences that creates a successful tourism development. One limiting factor, in many local government districts, is 'innovation competence' and this is usually found lacking in those areas with poor tourism developments. It is difficult to identify the resources that engender new competences.

The local government district of which we are speaking has good 'raw materials', a rich culture and a varied nature. It has a geographically strategic position in relation to Trondheim and Northern Møre. Both the Dovre railway and the E6 motorway go through this local government district. The tourist industry has a long tradition in this region and the local government district is surrounded by large protected natural areas such as Dovrefjell, Trollheimen and Forelhogna.

Action 1: a project to develop competence (1991–1995)

What were the results?

1. The principal goal of the project was to lay a foundation for the development of the tourist industry in the local government district. There was also a broadly based project to strengthen collaboration between the public sector, industry and landowners.

The strategy of action-based competence was chosen. The idea was to develop a project idea and to follow it through to its implementation. Within the project, we succeeded in bringing together 100 landowners to develop a common venture, the Children's World of Nature. It opened in 1995 and has functioned well until the present day. Indeed, it has won South Trøndelag's Open Air Activity Prize and has helped to build a positive reputation for the local government district. Furthermore, the project is an attraction that has won recognition in many parts of the country as well as internationally.

Through this project, the story or the *narrative* of 'Småjutulan' was told and explored. In a tourist context, the 'narrative' has the important function of constructing the identity of the experience, which is the product, creating a local sense of belonging.

The Children's World of Nature was justified by the following motives:

- a project that engages children to experience life in the open air;
- a meeting place for local people, the owners of holiday cottages, and other visitors;
- stimulus for collaboration between landowners via the development of a shared industrial activity; and
- an 'entry gate' to Trollheimen.

After 20 years in operation, the project is judged to have been very successful in relation to its aims.

2. Village associations were founded in several villages in the local government district and are still in operation. They received funds to develop their own local development plan.

3. An increased knowledge of tourism development. In retrospect, we see that the local government district and the villages have succeeded, to a limited extent, in making use of this knowledge.

Action 2: action-based development of competence is a study of guides to nature and culture

In a study context, the goal was to develop competences by developing ideas for projects or products that can be developed. If the aim of the projects is to be successful, much is demanded of these 'founders or zealous persons', which is contingent upon on receiving support from the public purse and on having a place in the local environment where they can work. A positive development of the tourist industry in a local government district will depend on identifying and making use of these ideas and projects as part of the further general development of the local government district.

Here the focus was on a study of guides to nature and culture in the region from Oppdal to Orkdal, for which the Norwegian School of Sports Sciences was responsible (1990–1991). The project tasks were given to the students which led to the development of many successful projects. Below are some examples of what happened in our local government district:

i. A project about local food at the annual village fair was launched. This development has continued as a successful product until the present day.

ii. A guide that describes the history, culture and cultural monuments was drawn up. This is available for those who visit the local government district.

iii. A local pilgrimage route was elaborated and implemented, the first in this local government district in modern times. Later, a national pilgrimage project was established; this Path goes through the local government district.

iv. Maintaining a postal station that existed in the local government district for a very long time.

Action 3: study of 'creation of values in industrial and cultural life': entrepreneurship in the school

This is what happened in the local government district:

1. Entrepreneurship in the school was initiated:

- a curriculum for 'entrepreneurship in the school' was created;
- four student essays were developed into platforms, which will be very helpful in continuing work on entrepreneurship in the school.

One part of the study was a planning day for teachers at the primary level, directed by students. Following the conclusion of the study, three planning days in entrepreneurship in the school were held at the primary and lower secondary levels.

2. Further development of the Pilgrim Path:

- The preliminary study on the Pilgrim Path, 'From Dovre to Nidaros Cathedral – from Church to Church', was carried out.
- The development of a 'pilgrim centre, 100 kilometres from Nidaros' was initiated. A preliminary plan for a shared building was drawn up, aiming at a collaboration between the school, the Normisjon missionary organization, the village association and the pilgrimage association.
- Four students wrote up potential projects providing a basis for increased industrial development.

3. A 'reputation process' was carried out as one part of the study. The county of South Trøndelag made a grant of 100,000 kroner available to support this process which enabled the establishment of a website (www.rennebu.com). This will be very important for the local government in the future. It displays the qualities of nature and culture in the locale, which will be highly significant for the tourist industry and for industry in general.

One consequence has been the continuation of projects such as the Children's World of Nature and The Journey to the Mountain. The local government has also provided a map of the area together with a project to elaborate signposts. These developments will be very important for advancing tourism within this region.

Retrospective reflections

What can be learned from this process? Whilst the project was active, many things went well, but we may wonder whether the improvement of competence has led to a lasting understanding of tourism development. We see over the course of time that the work on industrial development in the local government has returned to its old ways of marketing, administration, financing and leadership.

The process also shows, however, that against the background of a research-based development of the tourist industry, some actions have been carried out and plans for practical actions have been drawn up. This has created an understanding that if the competitive power in a region is to be increased, there needs to be a focus on the production of experiences and attractions.

If this is to succeed, a broad understanding of this insight must be established in the region. It takes time to establish stable competence functions and the development of attractions demands collaboration with organizations able to finance and construct the product. When attractions are developed, it is important to combine competences from several professional sectors which is vital if the quality of the product is to be ensured and maintained.

Within a local area the improvement of competence is extremely important in order to achieve a good result and this can best be developed by linking ideas and creativity to a real project(s). That is 'from idea to realization', which is an action-based development process of competence development.

Development of the Pilgrim Path in South Trøndelag into an Internationally Competent Tourist Industry Project: The Effects of Synergy on the Development of a Local Tourist Industry

Preliminary study of the Pilgrim Path

The preliminary study was financed by Innovation Norway in South Trøndelag. Its goal was to identify possibilities for a further development of the Pilgrim Path into an internationally competitive tourist industry project – a viable development in relation to nature and culture, the interests of society and financially benefiting those who supply products along the Path.

The preliminary study must be seen as a pilot project that can be very important for the development of more pilgrim paths in other parts of the country. The study was a contribution to a dialogue that seeks to generate a shared understanding of the Pilgrim Path as an international tourist product.

Our ancestors created the history and laid the foundations that make pilgrimages something that many people still want to experience. They created attractions by erecting beautiful church buildings and undertook the heavy investments that were involved. In comparison, only minor investments are required today in order to complete the Pilgrim Path as an international tourist industry project.

At present, it is up to state institutions to meet the expectations that they themselves have created. It is important to understand that we must focus on the infrastructure along the Path and implement a knowledge-based development. The Path must be well managed and meet the expectations of the pilgrims, involving the provision of overnight accommodation and the availability of food.

The preliminary study: the basis of the report

The initiative for this preliminary study was a consequence of a meeting arranged by the National Pilgrimage Centre in June 2012. The aim was to identify what had to be done in order to develop the Pilgrim Path from Dovre to Trondheim. Mayors, councillors and other representatives from Oppdal, Rennebu, Meldal, Orkdal, Skaun and Trondheim took part in the meeting; these were the local government districts supporting the preliminary study.

A dialogue with individuals and institutions involved in the work on the Path provided a platform to clarify various thoughts, ideas, goals, contrasts and presuppositions about its development into an international tourist project.

The development was discussed in connection with the following topics:

- What is the 'soul' in the Pilgrim Path?
- Today's situation
- The history up to the present day
- Developing competence

The report was based on conversations with pilgrims, suppliers and relevant officials in the Church and state bureaucracy who had been involved in work with pilgrims over a period of 20 years and was designed to be an open and transparent process. The application setting out the reasons for the preliminary study was sent to Innovation Norway and was communicated to the various groups involved; it was also presented in local newspapers so as to involve ordinary people, resulting in both positive and negative comments. The intention was to garner people's opinions and to discuss some of the more challenging situations that have been experienced in developing the Pilgrim Path. In the light of these comments, a memorandum was written in early July 2012 setting out the various points of view. This memorandum was presented and discussed with principal groups and persons who were working on the Pilgrim Path in South Trøndelag and was regularly updated.

Dialogue: What is the 'soul' in the pilgrimage?

For an international market, it is important to make the 'soul' of the pilgrimage visible. An essential aspect of this is meeting visitors' expectations through ensuring a sound organizational structure based on appropriate marketing and functional delivery of the product. For most of those on the pilgrimage route, it is a 'religious/spiritual' act, but just as important are the experiences of nature and of culture and participation in a living local environment. This was the conclusion of all the market surveys and questionnaires undertaken.

If the Path is structured in such a way that there is special overnight accommodation and good experiences of nature and culture, it could also be attractive to other groups of tourists. A quality-controlled Pilgrim Path could have important effects for the synergy of the local tourist industry.

The Pilgrim Path provides a number of important features that would appeal to an international market. These have been identified through questionnaires and the experience of tourist providers over the last 17 years:

- Nidaros Cathedral, the encounter with the tomb of Saint Olav.

'For me, Nidaros is about much more than just history; it is about identity and spirituality and it is both a gift and a challenge to people today' (Finn Wagle, Adresseavisen, 31 July 2012).

- Reflective walking; an authentic experience connected to religiosity/spirituality, nature, culture and what it means to be a pilgrim. The path to Nidaros Cathedral must lead to an intensification of the religious/spiritual expectations. This is why the tourist industry product Pilgrim Path must be marketed as stages between open churches. In South Trøndelag, from Dovrefjell to Nidaros Cathedral, from Eystein Church to Oppdal Church, from Oppdal Church to Rennebu Church, from Rennebu Church to Meldal Church, from Meldal Church to Skaun Church and from Skaun Church to Nidaros Cathedral.
- The churches are perhaps the finest cultural monuments along the Path, representing a living local milieu. This must be the core in the tourist industry product of the Pilgrim Path. The churches along the path must be the goal *en route*.
- The goal is walking.

Walking is, first and foremost, an experiential activity providing opportunities to enjoy nature, cultivated landscapes, fellowship and local communities. Walking also provides an opportunity to reflect on one's own life, on society's challenges in relation to nature and on one's fellow human beings, in both a local and a global perspective. Walking can also allow reflections on the past, the present and the future.

Bernd Lohse, a German priest who leads pilgrimages and has created great interest in Germany in our Pilgrim Path, has written a comment on the memorandum:

> I understand now that the pilgrim project comes from a milieu with deep roots, marked by the interplay between nature, culture, and religious and national identity. There is thus already an ecophilosophical background to the matter of pilgrimages. This is a fantastic framework with a high quality. There are ideas here that people throughout Europe need if they are to meet future challenges.

The ecophilosophical perspective has also been highlighted by Arne Bakken, formerly Pilgrimage Priest and Emeritus Bishop Finn Wagle:

> The pilgrim's day switches between walking, reflection and rest. We need to set up places that invite reflection *en route*, places of rest and shelter, where one can let one's thoughts drop to the ground.

There is also a need for special sleeping accommodation for those who seek 'rest for the night'.

Experiencing authenticity is a criterion of quality

For many hundreds of years, pilgrims have had memorable experiences along this Path. These include visiting old summer farms and farmyards, staying in fine mountain lodges, wooden huts, tourist stations, historic hotels and other places that have an important history. This information now requires development and updating for current pilgrims.

However, there is also the expectation to experience places that are representative of today's modern tourist facilities such as hotels and campsites encompassing new thinking, by developing modern architecture. In addition, there appears to be an increasing demand for participation in cultural events such as the pilgrimage days at Dovrefjell and the Kristin days in Skaun.

Today's situation

The critical question today relates to the return of tourist business and the investments made to provide tourist services on the Pilgrim Path. Quality sleeping accommodation is lacking on parts of the path and the total number of beds available is insufficient. The problem of capacity is exacerbated at peak times, especially during festival periods such as the Festival of St Olav in Trondheim at the end of July each year.

It is a challenge to develop low-cost accommodation in a high-cost country such as Norway, whilst at the same time ensuring high quality. The Pilgrim Project appears primarily to be a social phenomenon within a religious/value-oriented framework. Work on the pilgrimage route has created cultural values with many people believing that this was the real goal of this development. However, the Pilgrim Path has only, to a lesser extent, been a basis for the creation of a competitive tourist industry product. The preliminary study emphasized and discussed the possibilities of developing the economic potential of pilgrimages and perhaps a review of the current economic success is needed.

The history up to the present day

The interest in developing a Pilgrim Path to Nidaros increased at the end of the 1980s and the early 1990s. Several groups, working independently from each other with a variety of motivations, began working on preparing the Pilgrim Path from Oslo to Trondheim. This began in the first half of the 1990s, finally being realized in 1993. The Norwegian Directorate for Cultural Heritage was responsible for the development.

Five pilgrimage centres have been set up on the Path from South Trøndelag and a National Pilgrimage Centre was established in Trondheim. This was financed by the budget of the Norwegian Directorate for Cultural Heritage together with funds from the counties and the city of Trondheim. The Pilegrimsgård was founded as a centre for pilgrims. The 'living' pilgrimage work, administered by the pilgrimage association, is primarily centred along the long route that is taken on foot, from Oslo to Trondheim, with pilgrimage days arranged locally providing local walks for visitors. Furthermore, the length of the distance covered on foot is linked to the Festival of Saint Olav.

This part of the work on the pilgrim project has functioned well but it must be acknowledged that the resources that have been used to set up this organization have provided a product for a limited number of tourists/pilgrims. In view of the large investment in both money and labour it can be considered a very expensive venture, given the limited tourist monetary return. For example, a major challenge in Oppdal, Rennebu, Trondheim and other parishes has been to make the churches accessible to pilgrims throughout the entire pilgrimage season.

From a tourist economic perspective, up until now the work on the Pilgrim Path has produced very modest results. The challenge is to increase the number of walkers, which demands increased capacity in sleeping accommodation and in the supply of food. Security and the quality of people's experiences along the Path must also be improved.

It is possible for tourism and its supportive industries to profit from a comprehensive review of marketing that has been carried out. The Pilgrim Path in Scandinavia has been granted the status of a European cultural route called The Route of St Olav's Ways. It has thus become part of a larger collaborative project between the European Council and the EU with the intention of improving the effect that the status of a cultural route has on industry along the paths. Several projects have been advertised and St Olav's Way has been a partner in two large projects. This makes it one of three transnational routes among the 29 European Routes recognized by the Council of Europe, namely: Francigena Ways, Caminos de Santiago and St Olav Route.

One important part of the preliminary study was to designate a 'goal' for what could be done to further develop the Pilgrim Path from Dovre to Trondheim.

Developing competence

Developing competence is pivotal if an expansion in the tourist industry is to succeed. This is confirmed by the report *A Norway Based on Knowledge* (EKN) in 2012: 'For Norway, there remain two principal factors that are particularly important for industrial development and economic prosperity: natural resources and competence.' EKN concludes by presenting three overarching strategies that can give the tourist industries in Norway new growth and profitability. Some of these are quoted below:

> Industrialization. This means investing in effective logistics and standardization. Industrialization leads to increased volumes, lower costs and predictable quality, but it can also entail an increased burden on the environment.
>
> Self-service. This means that members of the workforce are replaced by self-service concepts, so that the elements of cost in industry are effectively reduced. The result is often simpler products, but this need not be at the cost of quality. One good example is the self-service mountain huts of the Norwegian Trekking Association.

> Unique experiences. This means developing niche products for unique or extreme experiences, often with a very high level of quality and service, at prices for an exclusive international market. These can be exclusive experiences of culture and nature.

EKN concludes that self-service is perhaps a more relevant strategy for Norway because of the level of wages. Sleeping in a hut for pilgrims with self-service has been evaluated and is deemed an authentic experience of high quality.

What can we learn from this advice? The pilgrim walks in order to have a unique experience of culture and nature and wants the walk to provide an opportunity for reflection. The pilgrim also has powerful experiences linked to 'places' *en route* – places that challenge or that are organized with this in view, unique places to rest, to have a meal and to take rest at night. The pilgrim pays close attention to the quality of sleeping accommodation, to the friendliness of the host, the quality of personal service and the cost of the services provided. Studies show that the cost of provision is extremely important for many people when choosing sleeping accommodation. They also prefer places that give them an authentic experience.

Experiencing high quality depends on the totality of the Pilgrim Path product. All of its elements must form a cohesive package; therefore, further development should focus on the quality of all the elements of the Path.

The County Governor in Oppland initiated central government involvement in the work on the Pilgrim Path in 1992. This led Torbjørn Berntsen, at the Department of Environmental Protection, to ask the Directorate for Cultural Heritage and the Directorate of the Administration of Nature to assist in its development. The project Pilgrim Path was classified as a developmental project from 1994 to 1997.

In the preparations for the Olympic Games (1994), an infrastructure of sleeping accommodation and high-quality experiences was set up in Gudbrandsdalen, which is still used by pilgrims today. A corresponding infrastructure has not been established between Dovre and Nidaros.

The biggest weakness in the Pilgrim Path project is the lack of infrastructure and quality control, which must be addressed if the Path is to become a competitive international tourist project. This particularly applies to parts of the Path in South Trøndelag.

Five pilgrim centres have been established south of Trøndelag, but none on stretches of the Path that could have the greatest potential for international development. Also, the pilgrim centre at Dovre has the task of carrying out quality control over the Path from Sel to Skaun, which is a long section, in relation to the total extent of the Path and the number of pilgrim centres.

A study trip to Santiago was conducted, intending to raise the competence of persons from the Church and the Norwegian state administration. Sarria, at 110 kilometres from Santiago de Compostela, is the most popular starting point for pilgrims walking to Santiago; indeed, 21.36% of all the 183,366 pilgrims who arrived at Santiago in 2011 began there. For Norway, this indicates that there is great potential in the last 100 kilometres, but also in the last *110* kilometres, if walking on Dovrefjell is presented as a great experience for pilgrims. The experience from Santiago ought to have led to the understanding that it was important to set up a pilgrim centre in Norway, 100 kilometres from Nidaros Cathedral.

Along the Pilgrim Path there are a number of locations to sleep and rest, linked primarily by walking through natural and cultural landscapes, both designed to give quality experiences to the pilgrim. However, some elements along the Path are absent, with others providing questionable quality. The key to improving this is redeveloping the infrastructure of the Path from Hjerkinn to Trondheim.

Up until now, large sums have been spent on reports, planning, meetings etc., which have all had the intention to raise competence and marketing opportunities for the Pilgrim Path. However, little has been spent on market-oriented product development in the form of infrastructure.

Retrospective reflections

This report proposes several necessary actions that must be implemented through a central pilgrimage administration, the county governor, the officers of the local government districts and suppliers along the Path. If these actions are not understood and accepted, it is hard to avoid seeing the inevitable consequences; namely, that the work on Norwegian pilgrimages faces an uncertain future as a commercially viable tourist project.

The potential for a positive development of the Pilgrim Path, 'From Dovrefjell to Nidaros Cathedral', is, however, great. It can become a very attractive international tourist product, greatly benefiting the tourist industry in Trøndelag.

A competence-based development of this section of the Pilgrim Path will also have a positive effect on the development of, and interest in, other pilgrim paths in Norway. Of prime importance for Norway and other pilgrim paths is 'walking to Nidaros Cathedral', which can act as a beacon for other international products in Norway.

If we are to succeed in elevating the Pilgrim Path to an international tourist product we must, first and foremost, be willing to re-evaluate the infrastructure of the pilgrim project, contributing new funds or redistributing the funds supplied by the public sector.

Conclusion

Both these cases underline the results of Sondre Svalastog's research. If competitive power in the tourist industry is to be increased, there must be a primary focus on the development of the product and the creation of structures enhancing the quality of experiences of nature and culture.

The Pilgrim Path has been successful and has been underpinned by academic theory. However, the delivery has focused on an aspect which has limited effect on encouraging competence in the tourism industry. Marketing, finance and leadership are seen as being the least effective for improving competitiveness in the international tourist market at a local level.

Although there are strong academic motivations behind the Pilgrim Path, we see that in this project, too, the focus is on the last link in the production process; first and foremost on marketing, administration, financing and leadership. However, this link has only a modest influence on the quality and adaptation to the market and thus on the competitive power of the industry.

References

Dahle, B. (1992) Notater og rapporter fra Norges idrettshøgskole, 106, *Hovedrapport*.

EKN (2012) *Destinasjon Norge Nasjonal strategi for reiselivsnæringen* Nærings- og handelsdepartementet, Norway. Available at: https://www.google.com/url?sa=t&source=web&rct=j&url=https://www.regjeringen.no/globalassets/upload/nhd/vedlegg/rapporter_2012/reiselivsstrategien_april2012.pdf&ved=2ahUKEwiuodHmt534AhVuQUEAHYVAAIUQFnoECBEQAQ&usg=AOvVaw1L_HnkFJd5EGokgKcELVHq (accessed 8 June 2022).

Svalastog, S. (2010) Om den raskt avtakende konkurransekraft i norsk reiseliv: en produksjonsteoretisk analyse. Research report 142/2010, Lillehammer, University College.

8 Norwegian Salmon Fishing Tourism: Only a Resource-based Tourism Product?

Øystein Aas, Stian Stensland and Sjur Baardsen

Introduction

Angling for Atlantic salmon (*Salmo salar* L.) emerged as one of the first significant forms of 'tourism' in Norway. Already in the 1820s–1830s, the first British travellers arrived, looking for rivers suitable for salmon angling (Berntsen, 1990; Gulbrandsen, 1994). Early pioneers who became established in Oslo, Bergen or Trondheim developed connections with civil servants, businessmen and academics who provided some information and then made their way through Norway, with ships, horses and carts. These early visitors, who certainly deserve the terms 'wilderness tourists' (Svalastog, 1994) or 'adventure tourists' (Buckley, 2006), were followed remarkably quickly by a significant number of other tourists. In just a few decades, this transformed salmon angling with rod, line and fly from a rare specialty of a few explorers into one of an expanding new tourism development (together with activities such as mountaineering, hunting and fjord cruises) of what Svalastog (1994) has labelled the Golden Age of Norwegian resource-based tourism between 1860 and 1914. Today, salmon fishing tourism is still a significant niche activity at municipal and private property levels around Norway, and other countries around the North Atlantic (Myrvold *et al.*, 2019). Annually, around 70,000 anglers – approximately 80% of whom are Norwegian and another 20% mostly Swedish, Danish, Finnish and German visitors – create expenditures related to salmon tourism in the range of 1260 million NOK[1]/year[2] (Stensland *et al.*, 2015; Andersen and Dervo, 2019). Today's salmon fishing tourism has changed significantly from what it was in the 'Golden Age' 100 years ago and it is difficult to picture a return to the former glory days. The purpose of this chapter is to present and discuss historical, recent and future salmon fishing tourism challenges in Norway, based on some of the key concepts that Sondre Svalastog presented, in an attempt to understand resource-based tourism. It raises a range of interesting questions about the history, the present status and the future development of resource-based tourism in Norway.

The Golden Age

The first British explorers started their visits in the 1820s and 1830s and the number of travellers rapidly increased, partly because several of the early travellers published books and reports about their adventures. Between 1860 and the outbreak of World War I in 1914, British salmon anglers completely dominated Norwegian salmon rivers. The famous tackle company Hardy of Alnwick, which also acted as an important travel agent, stated, in 1894, that 'All principal [Norwegian] salmon rivers are held by the British'. And by principal rivers, they did not

 Tourism as a Resource-based Industry: Based on the work of Sondre Svalastog (eds A.L. Svalastog, D.K. Müller and I. Jenkins)
DOI: 10.1079/9781800621480.0008

Box 8.1. Legal frames.

Most resources used in Norwegian nature-based tourism products are definitely *free goods*. The right to fish for salmon in Norway as well as in Scandinavia and the UK is, however, held by the owner of the land adjacent to the river (landowner) and therefore a private property right. This is the opposite of marine recreational fishing, in principle a public (free) good, as is also the general rule for salmon fishing in North America, in both fresh and salt water. While there may be regulations and/or fees to fish for marine fish and salmon in North America, the right to fish for salmon in private waters in Europe has been, and is still being, traded for significant sums. See Stensland (2011) for an overview of salmon fishing rights in Norway and the role of landowners in management and fishing tourism.

mean the limited number of famous rivers like the Vosso, Lærdal, Rauma, Gaula, Namsen and Alta, but also much less known rivers, even rivers that locals hardly consider worth fishing today. As an example, the very remote and quite insignificant Kongsfjordelva on the northern coast of the Varanger Peninsula was already used in the 1860s (Schulstad, 2009).

Most Norwegian rivers were extremely difficult to access and, at least in the beginning, could probably offer only poor accommodation and services. However, the main factor of production (the salmon) made nearly all salmon rivers in Norway highly competitive in the British market. Did they find pristine rivers congested with salmon? Actually, no, fishing was generally very good, but subsistence fishing with nets by locals was conducted on a large scale. Also, rivers were used for other purposes, in particular timber floating, which, in addition to subsistence fishing, created nuisance and conflicts with angling at certain locations and times (Gulbrandsen, 1994).

It is also likely that factors increasing demand played a major role in explaining this rapid development during the second half of the 19th century. This happened in the middle of the Victorian era, which was characterized by peace, growth and prosperity. The middle class grew significantly and during this period many modern sports and leisure activities were established or became common among the wider public in the UK. In Britain, recreational angling has its roots in the 1400s and has historically been divided into two distinct traditions: game fishing and coarse fishing (Lowerson, 1989). While coarse fishing (fishing for cyprinids, pike, perch etc.) has been associated with the working class, game fishing (fishing for salmon and trout) was a middle- and upper-class activity. By the mid-19th century, the salmon rivers of England, Wales, Ireland and Scotland were generally occupied and highly sought after, often by the upper class. Therefore, it has been suggested that lack of possibilities for the 'not so wealthy' upper class and an ambitious and fast-growing middle class was a major driver for the rapid expansion of Norwegian salmon fishing tourism between 1840 and 1880,[3] but we are not aware of any documentation that exists which could confirm this.

Soon, British visitors arranged a necessary touristic infrastructure (cf. Svalastog, 1998, p. 9). As early as 1848, *Jones' Guide to Norway* – essentially a salmon fishing holiday travel catalogue – was published for the first time and for several years this book came out with new annual editions. The travel agency Lumley & Dowell specialized in organizing fishing trips to Norway and was especially active in the latter decades of the Golden Era, often with Norwegian partners. Servants and 'gaffers'[4] were trained and hired and touristic superstructures, such as fishing lodges, were built. Guests brought along a range of British traditions with regard to angling, as well as modern cooking ideas, new house practices and general culture. Nevertheless, the localization factor of the Norwegian salmon fishing tourism must be assessed as a major drawback, especially compared to the short and easy trip to Scotland by rail, which took only a few hours, even in the 1800s.

The End of an Era

German submarines, which were deployed during World War I, abruptly halted the Golden Era (Sillanpää, 2002). The trip across the North Sea from England to Norway became too dangerous, especially when in 1917 Germany declared 'unrestricted submarine warfare' (openly targeting commercial ships, too). When the war ended in

1918, nothing would ever be the same again. Soon after the war, the world went into two economic depressions, with World War II starting in 1939. The period around 1920 also saw a temporary decrease in the salmon runs. The inter-war period still saw a number of British guests, especially those who had established strong material or emotional bonds with Norwegian districts. However, this period also saw the arrival of salmon anglers from other countries. The eccentric Nicholas Denissof signed his lease of the Aarøy in 1920 after an earlier lease held by Hardy of Alnwick. Denissof lived in France, after arriving as a refugee who had worked closely with the fallen Russian czar. His agreement lasted until the 1960s. Americans also became established in this period. William and Anna Singer first came to Nordfjord in 1913 and soon hired fishing in the Olden and nearby rivers. The Singer family made a fortune in the steel industry, while William Singer worked as a painter and lived off his inheritance. The Singers acted as early philanthropists, making huge efforts to improve the social and healthcare system in Nordfjord. This is just one of the many examples of how visiting anglers provided improvements and help to Norwegian districts during the 19th and early 20th centuries. They helped improve and modernize local agricultural practice, provided education for young people (sometimes abroad), offered positions and work, and provided support and capital to establish or improve public services.

The inter-war period saw Norwegian actors gradually playing a more significant role. Norwegians in general had, early on, become interested in salmon angling and the necessary skills and techniques were soon well established, not only in the upper social classes. When the ownership of Lord Portman in River Lærdal ended, the Norwegian shipping magnate Jørgen B. Stang took over in 1922 by buying Kapteinsgården. The family is still present. Well-established Norwegian companies such as Rieber, Beyer and Thams were active in the sale and provision of salmon angling to the foreign market, mostly acting as a Norwegian tour operator.

When World War II ended, rapid economic growth, combined with improved systems for general holidays and the expansion of private motoring, quickly established a significant and growing Scandinavian market for salmon angling tourism. Swedish and Danish middle-class segments became established on the bigger rivers such as the Gaula, Orkla, Surna, Namsen and Vefsna during the 1950s and 1960s. A large and growing number of Norwegian tourist anglers fished alongside local residents. In Finnmark, Finnish tourist anglers emerged as soon as the direct inland road connection to Finland was established in the 1960s. Only the most attractive and expensive rivers and beats remained as playgrounds for the market segments that 50–70 years earlier had dominated almost every river. Most of Norwegian salmon angling tourism has, since the 1970s, served a domestic Scandinavian market, comprising anglers from all classes of society. During the same decades, the resource situation has been highly variable and has included several incidents with a dramatic and often negative effect on the resource base, thereby also impacting on fishing opportunities and product quality.

Recent Trends and Challenges: Paradoxes of Travel Distance, Costs, Images and Reality of Product Quality

An attractive salmon angling tourism product of today is very similar to what it was 150 years ago. It includes access to an exclusive section of a river (or a whole river), a restricted number of anglers, a high probability of catching (large) salmon, decent accommodation, meals and also, maybe, guiding. In addition, transportation to and from as well as within the destination is needed. While Norway is considered the country with the largest number of healthy salmon rivers, several projects during the last two decades aiming to attract foreign, wealthy segments have changed the Scandinavian-dominated market only to a limited degree (Tangeland *et al.*, 2010; Stensland *et al.*, 2015; Aas *et al.*, 2021). Many rivers in Norway with a mediocre catch probability have tried to bypass low salmon returns by focusing on developing other aspects of their product such as facilities along the river, offering non-angling activities and targeting other groups, not simply those restricted to anglers. However, this has had limited success at least in attracting new and richer market segments, in line with Svalastog's framework which stresses the importance of the quality of the resource (salmon).

Box 8.2. The Atlantic salmon as a dynamic and unpredictable resource.

Salmon is often labelled the 'king of fish' and, as the most prized gamefish, it is sought by anglers from all over the world. The fish is anadromous, meaning they hatch in rivers where the juveniles live for the first two to four years, before migrating to the ocean where they grow fast, normally staying for one to three years. Angling takes place in rivers when the mature fish, usually weighing between 2 and 20 kg, migrate back to where they were born to reproduce. Like other biological resources, salmon productivity is dynamic and varies in space and time. At an early date, the desire to reduce variation and impose stability in salmon returns already sparked ideas about how to enhance salmon production. Artificial propagation and stocking had already started in the 1850s and fish ladders in waterfalls were constructed and in use from the early 1800s.

The species is naturally distributed around the North Atlantic from Maine in the USA via Canada, Iceland, north-western Russia, Scandinavia and, via the British Isles, into mainland Europe where it is distributed down to northern Spain. Industrialization from the 18th century onwards, with the damming, channelling and pollution of rivers, eradicated or greatly reduced the number of salmon in major European rivers, including the Thames and the Rhine. Today, its distribution and abundance are reduced, especially in the southern and urban/industrialized regions. Quebec and Labrador in eastern Canada, northern parts of the British Isles, Iceland, Norway and north-western Russia are now the only remaining regions with relatively strong populations. However, there are also regions and rivers within these areas that face problems. The largest number of rivers with viable populations are currently found in Norway and Canada.

In Norway, salmon issues have been embedded in laws since the Middle Ages. In particular, concerns about over-harvesting, distribution and fairness between sea and river harvest, as well as between holders of rights in a river, have been addressed. From the start of the 20th century, the development of hydropower, as well as the unexplained death of salmon in rivers in southern Norway, represented the first signs of systematic reductions in Norwegian salmon stocks. During the 1960s, ocean and marine fishing for salmon grew rapidly at the cost of inshore and river fishing, before this was regulated by international laws such as the NASCO treaty and the establishment of the Law of the Seas during the 1980s. Despite the introduction of more sustainable regulations, salmon stocks have continued on a downward spiral. During the last three decades, the survival of salmon at sea has fallen considerably, leaving a much smaller surplus available for fishing. The development of salmon aquaculture and the global demand for farmed salmon have consequences far beyond the aquaculture site. These operations impose new challenges and negative impacts on stocks and salmon fisheries due to parasites, disease and genetic and ecological interactions (Forseth *et al.*, 2017). (See Aas *et al.*, 2011 for an in-depth presentation of salmon ecology.)

Unlike the 1850s, when visiting anglers had to spend several days on often rather uncomfortable journeys across the North Sea to reach Norway, UK anglers nowadays can jump on a low-cost flight from one of the many airports in the UK and be in Trondheim or Alta in just a few hours, often on a direct flight. The drive from Trondheim airport to nearby salmon rivers takes between 15 minutes and two hours. Paradoxically, the distance between the UK and Norway seems longer than before despite the huge ease, and reduction in time and cost, of travel. In total, only a few hundred visitors from the UK visit Norway to fish for salmon nowadays. Why is this? The answer can be analysed from at least three different perspectives.

Market Trends and Competing Destinations

One of the main arguments put forward by Svalastog (2006) as to why the Golden Age of Norwegian tourism is over is because Norway has become a high-cost country. Competitiveness in leisure tourism is linked to both the price and the quality of the product, where quality is decided primarily by two factors: quality of the raw material (i.e. salmon) and the ability to gain and utilize skills and knowledge (e.g. things about the fishing product other than the abundance of salmon) (Svalastog, 1998, p. 11). We would therefore argue, in line with Svalastog (2006), that there are other factors in the industry and the region itself that determine whether the salmon angling product is, in the long run, competitive. Salmon angling has never been, and never will be, a product for the masses. It is a specialist niche product that, historically, has been expensive, targeting segments that are 'well off' or have such a strong interest that they give it high priority and spend a relatively large amount of their income on it.

Major destinations for high-quality salmon angling today are all the Icelandic rivers with their lodges, most rivers on the Kola Peninsula in Russia (remote lodges but also tented camps), together with some rivers, beats and lodges in Canada, Scotland and Norway. The cost for one person varies from more than 10,000 Euro[5] per week for the best products, down to 1000 Euro for the cheaper products. The basic ingredients of salmon angling products are, supposedly, high-quality fishing (restricted to the group) and at least self-catering accommodation. Meals and guiding are included in some products without greatly affecting the total price. Most Norwegian products are in the lower price range from 1000 to 3000 Euro, except for the Alta River, which probably has the costliest and most sought-after salmon angling product in the world.

Although the domestic market is by far the most important for providers of salmon angling in Norway, more and more Norwegian salmon anglers travel abroad to fish for salmon. This was something almost unheard of 20 years ago. Norwegian visitors are now the second largest nationality among salmon anglers in the Scottish river Dee, one of the four most important rivers and a relevant indicator for all of Scotland (Ken Reid, manager, FishDee, personal communication, June 2013). Only domestic (UK) anglers make up a larger segment. Here, the revolution in airline route offers and lowered prices, combined with the relatively high salaries and the large percentage of dedicated salmon anglers in Norway, are likely important reasons for this. Interestingly, while destinations offering salmon angling are often competitors, the products in Scotland and Norway are less so, since the prime seasons differ. While the peak seasons in Scotland are spring and autumn, Norway has its top season during the summer. Nevertheless, it is intriguing how Norwegian anglers have taken advantage of the reduced costs of medium- and long-distance travel. However, we see few or no signs of the same happening in Norwegian salmon-angling tourism, at least in terms of an increase in market segments originating outside the Nordic countries. While the high-cost level in Norway might explain why mainstream market segments from abroad have been avoiding Norwegian tourism products, we would not expect this to be a major factor in the specialized salmon tourism market.

Images and Realities of Norwegian Salmon Angling Products

Salmon stocks have been under pressure and on a downward trend for several decades, most of all in the southern regions of its original distribution. Although Norway is one of the countries with the largest remaining salmon returns, the country has experienced many challenges and negative impacts on its salmon stocks. The spread of the lethal parasite *Gyrodactylus salaris* in the 1980s and 1990s has been disastrous. This parasite hit stocks very hard in several iconic rivers such as Rauma, Driva, Lærdal and Vefsna and, in practice, ruined the fishing tourism for salmon for many decades, although the Vefsna, Rauma and Lærdal now have partly recovered and attract fishing tourists again. Other high-profile rivers such as Aurland, Eira and Suldal have experienced detrimental impacts from hydro power regulation; this is in addition to general impacts such as a high harvest rate both in the sea and in rivers, poor marine conditions and low survival, the high number of escaped farmed salmon,[6] and post-smolt mortality induced by sea lice. In addition, social and cultural aspects of Norwegian salmon angling differ from those in the UK and Atlantic North America (Myrvold *et al.*, 2019). In Norway, salmon angling became an activity for all, with the possibility of using many types of gear and harvesting salmon at a low cost, often in unregulated or poorly managed rivers. This domestic practice did not mix well with high-end angling products (Øian *et al.*, 2017).

All of this seems to create an image that everything, or nearly everything, about Norwegian wild salmon is poor including salmon conservation and angling products (Millington-Drake, 2000). As Svalastog (1998) points out, neither the quality of the resource nor the quality of knowledge is sufficient to make a competitive product. Also, improvements for salmon, such as the highly successful restoration of salmon in many rivers in southern Norway, are disregarded by those who have a negative image of Norway's salmon management. However, a few rivers, especially the Alta, but also the Gaula and Orkla, have upheld a positive though vulnerable reputation over time and, more recently, rivers like the Lakselva and Vefsna in northern Norway have received positive feedback, especially for the chance to catch very large trophy fish. Interestingly, a dramatic drop of about 12% in active salmon anglers in Norway from 2012 to 2013 coincided with a significant decrease in the salmon runs in the Gaula and Orkla, the two rivers which during the last decade have probably been the most visited Atlantic salmon rivers in the world in terms of number of rod-days.

Supply and Product Development

What positions and what strategies are providers taking? Do actors have the necessary amount and level of knowledge (competence and skills) to produce a competitive salmon angling product? Fishing rights are generally owned by landowners with a farm property. This group is therefore the main supplier of angling in Norway. Many landowners also provide accommodation, meal service and other services such as guiding. The landowner group is, however, heterogeneous, with great variations in landowner and property characteristics (Stensland, 2010, 2012). This includes length and catch possibilities on their fishing stretch, the number of rods allowed, services provided and actual and potential income from salmon angling tourism. Average net income from salmon angling in the four major Trondheim fjord salmon rivers, Gaula, Orkla, Stjørdal and Verdal, was 29,000 NOK in 2008, but the variation was large and one in three had no income (Stensland, 2010). Landowners choose whether they want to keep fishing for themselves, rent out to fishing tourism entrepreneurs/other landowners, or deal with the anglers directly. The rational landowner maximizes his utility by receiving equal marginal utility from all his potential income sources (Johansson and Löfgren, 1985). The structural rationalization and diminishing margins in the agriculture sector means that fewer landowners work on the farm, resulting in less income from farm resources. This also changes their relationship with farm resources, the implication being that recreational or even no objectives about the farm and fishing rights are gaining priority over economic objectives (Stensland, 2012). For most, salmon angling tourism is not a high priority strategy to secure household income (Stensland, 2013; Stensland and Kristiansen, 2013). Unfortunate effects of this could be less supply and lower product quality. Landowners could choose not to offer fishing on an open market but keep fishing for themselves or simply not care much about renting out at all, because there is not much money to be made (Stensland, 2012). Stensland and Baardsen (2012) found that marginalization of salmon angling tourism income (compared to other income sources) decreases profit efficiency, so that the landowner is wasting his resources by obtaining less profit than the maximum obtainable profit. They argue that, from an economic sustainability perspective, this is unfortunate and call for growth of more professional fishing tourism entrepreneurs with a larger portfolio of fishing as a way of counteracting this. The average landowner simply does not have sufficient fishing tourism competence, salmon fishing interest, or quality of the fishing right that would be required to develop a prime competitive angling product. As Svalastog (2006) would have argued, the imbalance between the development of knowledge and the relative growth in salaries and costs between sectors has helped to weaken the competiveness of Norwegian salmon fishing products. Knowledge-building among fishing tourism providers can, however, be facilitated. A professionalization of fishing tourism suppliers and administration of fishing stretches done by a few skilled fishing tourism entrepreneurs might be a key strategy for growth. Current legislation does not allow land ownership and fishing rights to be split. Unlike other farm resources, for example agricultural fields, the fishing right can be rented out for a maximum of ten consecutive years. Stensland (2013) suggests introducing a minimum size for beats (fishing stretches) to be open for fishing, encouraging more fishing beats to be available for the market. Landowners are already familiar with such an arrangement from the managing of moose-hunting quotas. Iceland has an interesting system whereby landowners are forced to lump their beats into one 'pool' and the whole river, or larger stretches of it, is leased to one professional operator or an angling club (Stensland *et al.*, 2017a). Fishing in Iceland is expensive and has, typically, considerably fewer rods per kilometre of river than in Norway, thereby increasing the catch success of anglers. Changing objectives about farm resources and current legislation make it harder for some individuals to buy or lease fishing rights and thereby build the knowledge required to specialize in angling tourism, as suggested by Stensland (2013).

A Future? Focus on Resilience, Responsibility, Knowledge and Quality

Salmon angling in Norwegian rivers has recently been diversified. The harvest orientation, formerly strong, is perhaps the motivation factor that has changed most significantly over the last 10–15 years. While subsistence values such as 'Do not play with the food', with a reference to catch and release (C&R), have been dominant in Scandinavian recreational fisheries (Aas *et al.*, 2002), we are now witnessing a rapid change. National catch-and-release statistics for salmon rivers were first recorded in 2009, with 7% of all anadromous fish (salmon, sea-run char, trout) being released. In 2020, the release rate had grown to 23% (Statistics Norway, 2021) Public opinion about C&R was, in 2012–2017 for the first time, positive, while by 2020 it had changed with a slight majority having a negative attitude to releasing fish (Kantar, 2020). However, the registered C&R rate in Norwegian rivers varies from zero to more than 80%, suggesting that different rivers have their own breed of anglers. A study from Lakselva river in northern Norway suggests that a norm for C&R is emerging and growing among anglers. Stock concern, in particular, but also social pressure from fellow anglers, are important drivers for this growing C&R norm (Stensland *et al.*, 2013). Several studies show both the existence of a norm for keeping fish and one for releasing fish, suggesting a possible polarization of angler behaviour and opinions on the C&R issue (Stensland and Aas, 2014; Øian *et al.* 2017).

Changes in stock abundance, further catch restrictions and the emergence of a C&R norm have implications for the type and amount of angling products offered, and also how much anglers fish (Stensland *et al.*, 2017b). C&R is likely to grow, while keeping fish is decreasing, but still happening. Under such a scenario, with tight harvest restrictions, Stensland *et al.* (2021) find that the recreation value of anglers decreases, and suggest that, over time, older harvest-oriented anglers will be replaced by younger fly-fishing C&R anglers. Skullerud and Stensland (2013) argued that the two growing angler groups – the 'trophy angler' and the 'C&R angler' – are the two market segments best suited for the future since they spend more money than others and are less harvest-oriented than other segments.

The natural fluctuation in salmon abundance over the years also makes salmon angling tourism a risky business. Recent changes and restrictions in Norwegian salmon fisheries from 2009 onwards have clarified this further. Fishery issues, including a shortened season, variations in stock abundance and angling demand, have been perceived as the top risk sources impacting on landowner income from salmon angling tourism (Stensland, 2013). Perceived risk is a subjective matter affecting the economic environment, and thus economic decisions by individual landowners. With increased risk, fewer landowners will attempt to make an income from salmon angling tourism and development is not happening.

It seems obvious that ensuring long-term viability, survival and abundance of salmon is a prerequisite for development of sustainable angling tourism. Maximizing natural smolt production in the rivers through harvest management, habitat management and habitat improvement are important conservation measures, even if many landowners (Stensland, 2012) perceive the major problems to be elsewhere and caused by persons other than the stakeholders in the rivers. Climate change could further cause challenges to evolutionary adaptations and future survival of salmon stocks (Todd *et al.*, 2011). Expanding habitat opportunities in freshwater, so that salmon stocks can express their maximum evolutionary life-history variation and better respond to climate changes, will strengthen resilience and survival of the stocks and is therefore crucial (Bottom *et al.*, 2009). Landowners are key actors in this regard since they are owners of salmon habitat and managers of the stocks manifested through their collective work on these fishing beats. Fundamental institutional challenges remain since major externalities (costs) are not sufficiently internalized, so that losses created by other sectors allow these businesses to act as free riders.

Acknowledgements

We would like to extend our thanks to the reviewers and the editors of this anthology. This paper is part of the project SALMONCHANGE, funded by the Research Council of Norway (project no. 208056), the Norwegian Environment Agency, the Agricultural Research Fund and Porsanger Development Fund.

Notes

[1] Euro = 10.00 NOK as of 1 October 2021.
[2] A detailed study of the economic impact of salmon fishing in the Trondheim region, examining primarily the rivers Orkla, Gaula, Stjørdal and Verdal, in 2008 estimated expenditures in the range of 227–254 million NOK, while the local economic impact was calculated at 82–106 mNOK (Fiske *et al.*, 2012).
[3] This is, to some extent, in opposition to the often-repeated statement that all or most British visiting anglers were 'lords'. However, even if the lords did not dominate, there were certainly lords among them. Lord Bromley-Davenport came to Rauma and the Duke of Roxburghe came to Alta early on. In both cases, their descendants still fish these rivers.
[4] The Norwegian term 'klepper' is synonymous with 'gaffer', which means the man who used a 'gaff' (a large hook on a long stick) to land the fish when it was brought to shore. Today, especially with the growth of catch-and-release fishing, the term is less used and many would find it highly inappropriate. Instead, terms such as 'gillie' (helper of angler/hunter) and guide are more common. Other more local terms are used such as 'staker' ('poler') for a guide/boatman on certain large rivers in northern Norway.
[5] While Norway in general is struggling with providing high-end products, the most expensive salmon angling product in the world is most likely the exclusive, syndicate-organized fishing on the river Alta, northern Norway. The cost for one week's fishing for one angler is up to 300,000 NOK (Leinan, 2007).
[6] While escaped farmed salmon may be caught by anglers, most anglers perceive the catch of escapees as something that detracts from the angling experience. This makes them less willing to pay for recreational fishing than for fishing for a purely wild stock (Olaussen and Liu, 2011).

References

Aas, Ø., Thailing, C.E. and Ditton, R.B. (2002) Controversy over catch-and-relase recreational fishing in Europe. In: Pitcher, T.J. and Hollingworth, C.E. (eds) *Recreational Fisheries: Ecological, Economic and Social Evaluation*. Blackwell, Oxford, UK, pp. 96–106.

Aas, Ø., Einum, S., Klemetsen, A. and Skurdal, J. (2011) *Atlantic Salmon Ecology*. Wiley-Blackwell, Oxford, UK.

Aas, Ø., Andersen, O. and Stensland, S. (2021) *Laksefiske i Norge 2020: Atferd blant aktive fiskere og svenske fiskere som ikke besøkte Norge pga. corona-pandemien*. NINA fagrapport 7. NINA, Lillehammer, Norway.

Andersen, O. and Dervo, B.K. (2019) *Jegernes og fiskernes forbruk av varer og tjenester i Norge i 2018*. NINA Rapport 1605. Norsk institutt for naturforskning, Lillehammer, Norway.

Berntsen, B. (1990). De dansende fluer. Oslo, Grøndahl, p. 197.

Bottom, D.L., Jones, K.K., Simenstad, C.A. and Smith, C.L. (2009) Reconnecting social and ecological resilience in salmon ecosystems. *Ecology & Society* 14(1).

Buckley, R. (2006) Adventure tourism. The new frontier.

Fiske, P., Baardsen, S., Stensland, S., Hvidsten, N.A. and Aas, Ø. (2012) *Sluttrapport og evaluering av oppleieordningen i Trondheimsfjorden*. Korrigert versjon av NINA Rapport 546. NINA, Lillehammer, Norway.

Forseth, T., Barlaup, B.T., Finstad, B., Fiske, P., Gjøsæter, H. *et al.* (2017) The major threats to Atlantic salmon in Norway. *ICES Journal of Marine Science* 74(6), 1496–1513.

Gulbrandsen, P.G. (1994) Med karjol og laksestenger for 150 år siden. Espa: Lokalhistorisk forlag.

Johansson, P.-O., and Löfgren, K.-G. (1985) *The Economics of Forestry and Natural Resources*. Basil Blackwell, Oxford.

Kantar (2020) Natur- og miljøbarometeret. Nordmenns holdninger og atferd i natur- og miljøvernspørsmål. (Available at: norskfriluftsliv.no/wp-content/uploads/2020/11/Natur-og-miljobarometeret-2020-31.10.2020.pdf (accessed 1 October 2021).

Leinan, I. (2007) Laksefiske i Altaelva. Innlegg på konferanse om laksefisketurisme.

Lowerson, J. (1989) Angling. pp. 12–44 in Mason, T. (ed). *Sport in Britain - a social history*. University of Cambridge Press, Cambridge.

Millington-Drake, T. (2002) International market trends in recreational salmon fisheries – consequences for Norway. In S. Stensland (ed.), Fisketurismekonferansen Gammel næring –ny giv. Foredrag. Rapport (pp. 31–36). Oslo, Norway: Norske Lakseelver.

Myrvold, K.M., Mawle, G.W., Andersen, O. and Aas, Ø. (2019) *The Social, Economic and Cultural Values of Wild Atlantic Salmon: A Review of Literature for the Period 2009-2019 and an Assessment of Changes in Values. NINA Report 1668*. Norwegian Institute for Nature Research, Lillehammer, Norway.

Øian, H., Aas, Ø., Skår, M., Andersen, O. and Stensland, S. (2017) Rhetoric and hegemony in consumptive wildlife tourism: polarizing sustainability discourses among angling tourism stakeholders. *Journal of Sustainable Tourism* 25(11), 1547–1562.

Olaussen, J.O. and Liu, Y. (2011) On the willingness-to-pay for recreational fishing: escaped versus wild Atlantic salmon. *Aquaculture Economics & Management* 15(4), 245–261.

Schulstad, T. (2009) *Historien om Kongsfjordelva*. Berlevåg JFF/eget forlag, Berlevåg.

Sillanpää, P. (2002) *The Scandinavian Sporting Tour. A Case Study in Geographical Imagology*. Östersund, ETOUR, Vetenskapeliga Bokserien 9.

Skullerud, E. and Stensland, S. (2013) Laksefiskere og motivasjon for å fiske i Lakselva, Finnmark. *Utmark* 2013(1).

Statistics Norway (2021) River catch of salmon, sea trout, migratory char, rainbow trout and pink salmon, 2020. Available at: www.ssb.no/en/jord-skog-jakt-og-fiskeri/statistikker/elvefiske/aar (accessed 1 October 2021).

Stensland, S. (2010) Fishing rights and supply of salmon angling tourism in mid-Norway. *Scandinavian Journal of Hospitality and Tourism* 10(3), 207–230.

Stensland, S. (2011) Angling tourism and landowner based management of Atlantic salmon stocks in the Trondheim fjord region of Norway. PhD Thesis no. 2011:26, Norwegian University of Life Sciences. http://statisk.umb.no/ina/forskning/drgrader/2011-Stensland.pdf

Stensland, S. (2012) Typology of landowners in Norwegian salmon angling: attitudes towards river owner organisations and management actions. *Fisheries Management and Ecology* 19(4), 273–282.

Stensland, S. (2013) Landowners' perception of risk sources and risk management strategies in Norwegian salmon angling tourism. *Scandinavian Journal of Hospitality and Tourism* 13(3), 208–227.

Stensland, S. and Baardsen, S. (2012) The effects of property and landowner characteristics on profit efficiency in salmon angling tourism in Norway. *Journal of Sustainable Tourism* 20(4), 627–644.

Stensland, S. and Kristiansen, A.H. (2013) Elveeieres målsettinger, framtidssyn og inntekststrategier i en laksefisketurisme i endring. *Utmark* 2013(2).

Stensland, S., Aas, Ø. and Mehmetoglu, M. (2013) The influence of norms and consequences on voluntary catch & release angling behavior. *Human Dimensions of Wildlife* 18, 373–385.

Stensland, S. and Aas, Ø. (2014) The role of social norms and informal sanctions in catch-and-release angling. *Fisheries Management and Ecology*. doi: 10.1111/fme.12078.

Stensland, S., Fossgard, K., Andersen, O. and Aas, Ø. (2015) *Laksefiske i endring. – En spørreundersøkelse blant sportsfiskere som drev elvefiske etter laks, sjøørret og sjørøye i Norge 2012–2014. Norwegian University of Life Sciences*, Ås, Norway.

Stensland, S., Agnarsson, S., Helgason, T., Jóhannesson, G.T., Larsen, F. and Aas, Ø. (2017a) *A Survey of Resident Anglers Fishing for Salmon, Trout and Char in Iceland*. Norwegian University of Life Sciences, Ås, Norway.

Stensland, S., Aas, Ø. and Mehmetoglu, M. (2017b) Understanding constraints and facilitators to salmon angling participation: insights from structural equation modeling. *Human Dimensions of Wildlife* 22(1), 1–17.

Stensland, S., Dugstad, A. and Navrud, S. (2021) The recreational value of Atlantic salmon angling under different fishing regulations. *Fisheries Management and Ecology*. DOI:10.1111/fme.12487.

Svalastog, S. (1994) Lokalisering av reiseliv. Om ressursanalyser, den romlige fordeling og lokal innpassing. Skriftserien nr. 92/1994. Lillehammer: Høgskolen i Lillehammer.

Svalastog, S. (1998) The nature of resources for tourism and recreation. Working paper no. 68/1998. Lillehammer: Høgskolen i Lillehammer.

Svalastog, S. (2006) Noen kommentarer til fenomenene: ressurser, identifisering av ressurser, og kvalitet. Unpublished note of 25 Aug 2006. p. 5.

Tangeland, T., Andersen, O., Aas, Ø. and Fiske, P. (2010) *Elvefiske etter anadrome laksefisk i Norge sesongen 2008. Fiskevaner, fangst, innsats og holdninger til fangstreguleringer, herunder fiskernes syn på reguleringene for sesongen 2008*. NINA Rapport, 545. Norsk institutt for naturforskning (NINA), Lillehammer, Norway.

Todd, C.D., Friedland, K.D., MacLean, J.C., Hazon, N. and Jensen, A.J. (2011) Getting into hot water? Atlantic salmon responses to climate change in freshwater and marine environments. In: Aas, Ø., Einum, S., Klemetsen, A. and Skurdal, J. (eds) *Atlantic Salmon Ecology*. Wiley-Blackwell, Oxford, UK, pp. 409–443.

9 Post-modern Niches in Tourism Development

Ian Jenkins

Introduction

This chapter illustrates how economic restructuring affects post-industrial economies; in particular, the move from a modernist/Fordist dictum to that of a post-modern society, which appears associated with a diversity of products and the de-massification of markets (Urry, 2002; Papathanassis, 2011). This critical change seems to promote and give agency to the concept of economic niches, incorporating a transformation from standardized mass-produced products and limited consumer choice to a market that exhibits an array of product diversity (Ritzer, 1996; Sert, 2017; Rossetti and Quinn, 2019). The concept of 'globalization' appears especially pertinent to tourism, the essence of which is based upon space and time. This compression, according to several commentators, seems to be driving globalization (Ritzer, 1996; Shaw and Williams, 2002; Pender and Sharpley, 2005).

This chapter will demonstrate that resource allocation and tourism product development are important for ensuring that countries remain competitive in a global market. It explores the concept of a niche in the context of the tourist market and how the niche became a strategic tool for encouraging tourism to a particular destination. It emphasizes Svalastog's treatise that resource management development is more than just a physical element but also incorporates cultural and human aspects. These are clearly necessary to develop a niche tourism product at a successful tourist destination. Niches use very specific and specialized resources (physical, cultural/human) to assemble their products. Many are distinctive types of tourist niches, often developed from more general segments of the tourism market, such as cultural tourism (Hoppen *et al.*, 2014). The later part of the chapter uses the exemplar of a tourist niche (Jenkins, 2019) to demonstrate how resource management of a destination (identified by Svalastog) is able to develop a niche product. The exemplar is the Hay Festival, which can be considered to be a Hallmark Festival and is one of the most successful literary festivals (VisitWales, 2021), contributing to the sustainability of a small Welsh town. Kotler confirms the concept of niches as being part of segments (Kotler, 2000). This further supports the development of smaller markets that may be satisfied by one or a few specialized products, which could be considered to be niches (Marson, 2011; Sert, 2017). There has also been a rise in the importance of niches (Papathanassis, 2011); a review of the commentary on these demonstrates no singular, definitive concept but rather the emergence of a number of different types of niches. Consequently, there is a need to develop a model typology of all current understandings of niche tourism. Figure 9.1 summarizes these elements and a succinct explanation

DOI: 10.1079/9781800621480.0009

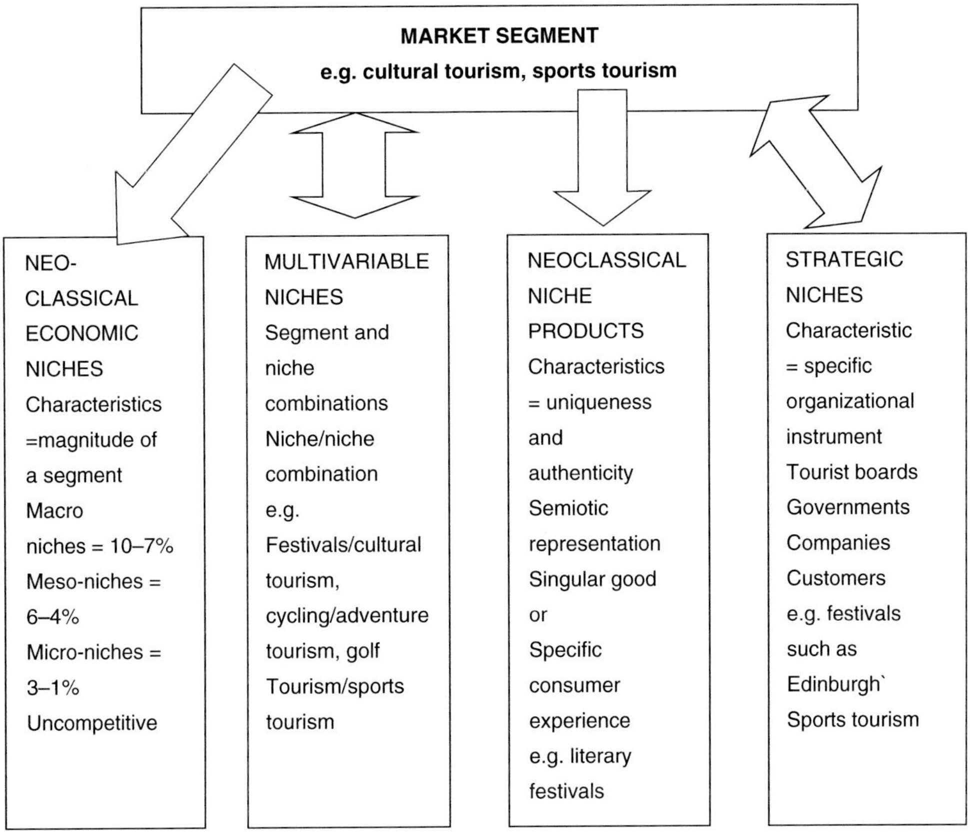

Fig. 9.1. Typology of tourism niches.

of each is found below, coupled with a more detailed description and justification of their typology.

The **neoclassical market niche** is postulated as having economic viability and is representative of a small section (magnitude) of a market segment. These niches can be further subdivided by magnitude (market size) into the three elements of micro, meso and macro, reflecting proportional differences to market segments, possibly 7–10% being macro (BTA, 2002), 4–6% meso and 1–3% below. Micro is categorized as a very small niche, representative of, for example, UK literary festivals. This latter category may also include the notion of an uncompetitive market, more akin to the original biological definition of a niche.

The justification for this category can be supported by National Tourism Organizations such as Tourism Australia (2005), DCMS (1999), BTA (2002) and Bord Fáilte (O'Reilly, 1999) who describe niches as having a definable magnitude related to a percentage of a segment, with upper limits varying from 7% to 10%. It is also concomitantly connected to the notion of post-modernism whereby markets are constantly differentiating into smaller sections. MacLellan also concurs with the notion of a niche being an array of elements such as macro and micro parts of a segment (MacLellan, 2003). Consequently, it seems logical to divide a niche into various sizes and the concept of both macro and micro niches is supported by authors such as MacLellan (2003). The meso niche division is a further adaption (Jenkins, 2007) and a logical division between the other two categories.

A **neoclassical product niche** is theorized to exhibit specific and limited products, offering unique and authentic objectives or experiences for the consumer; the antithesis of mass-tourism products; for example, a Shackleton and Scott Cruise of Antarctica (Adventure Associates, 2005).

The validation for this can be taken from authors such as Choat (2004), in which it is the nature of the product which is important, rather than the notion of economic markets. This theme is also supported by various authors such as Doyle (2003) and Page *et al.* (2001), who cite specific and unique products as differentiating these from segments, providing evidence of niche creation. Further verification is provided by some of the UK tourist boards, who identify specific products as examples of niches (WTB, 2004; STB, 2005). Evidently, VisitEngland, as cited by Mayfield (interviewed in 2005), sees British 'movie maps' as a hook for tourist visits. Further confirmation and evidence for the existence of these niches is represented by the private sector, such as specific cruises associated with explorers (Adventure Associates, 2005). Urry (2002), and possibly Baudrillard (1999), also see the new specialized products as being 'semiotic' rather than 'utilitarian', giving these new niche products a specialist and cultural cachet different to that of mass tourism.

Multi-variable niches are postulated to be segments which are used by consumers/tourists to facilitate access to more specific products, similar to those identified in neoclassical products. For example, a trip to Australia is considered to be a segment including visits to nature reserves and Aboriginal culture (Tourism Australia, 2005), which are regarded as niches.

Evidence for the existence of this type of niche is profferred by some NTOs, for instance Tourism Australia (2005) and, to some extent, by BTA (2002). They both cite the consumption of niche products as only a part of the holiday experience and not necessarily the main purpose of the trip. Further primary evidence is garnered from Brinkworth (interviewed in July 2002), who also suggests that niche products are rarely the sole reason for an international visit, advocating that visitors may consume both niches and segments whilst visiting a destination.

Strategic niches are based on organizations identifying niches as opportunities to develop marketing strategies to increase consumption of products. Certain markets and products are used as strategic instruments to focus marketing and are created from specific consumer profiles. This can be akin to the DCMS's strategy (1999) for UK tourism. However, the DCMS tourism report of 2011 makes little reference to niches (DCMS, 2011).

Justification for this type of niche is inferential and vicarious. The DCMS certainly decided that niches could be used for strategic purposes and in 1999 stated that they were one element of a UK strategy (DCMS, 1999). Nonetheless, by 2003, niches were not directly featured (DCMS, 2003). A number of NTOs' strategic documents isolate elements of segments, such as golf and literary tourism, for example Bord Fáltie (O'Reilly, 1999), VisitScotland (The Scottish Parliament, 2002) and Visit Wales (WTB, 2002). However, other tourist boards such as VisitEngland do not directly identify niches. Nevertheless, certain regional tourist boards cite niches as important, for example the south-west of England (Page *et al.*, 2001). Further evidence for the existence of strategic niches is garnered from some authors including Page *et al.* (2001) and Dahringer and Muhlbacher (1991) who note that niches might be used in new ways to create destination features. Therefore, niches may have a strategic part to play in the development of tourist destinations.

Nonetheless, what is apparent is that niches, like other tourist products, are based on the theory of resource allocations as defined by Svalastog (see Chapters 1–5, this volume). It is also interesting to note that geography plays a key role in understanding tourist resources. The author is a geographer and would support the theory assertion that developments of tourist resources are based upon, firstly, location factors and, secondly, their use of resource factors, both tangible and intangible. What is also evident is that many tourist resources are, in fact, intangible (Svalastog, see Chapters 1–5, this volume) and therefore difficult to evaluate. However, it seems that many current 'niche tourists' wish to participate in experiential products within the area of resource allocation, in which many resources are intangible. Although location is a key tourist component for motivation to travel, the actual experiential aspect of the tourist trip is clearly very important for certain types of niches. Certainly, this is the case in Cohen's and Plog's typologies, especially those at the more 'adventurous' ends of the tourist spectrum such as explorers and psychocentrics. This is reflected by many of the new tourist niches which seem to specialize in 'experience' of the product rather than a physical product *per se*. For example, the growth of adventure tourism, festivals and events can be explained to some degree in this way, as many of these products relate to the classic understanding of niches. It is also manifest that most tourist products are, in effect, related to perception and service, which are, in essence, difficult to qualify and quantify.

Festivals, and in particular literary festivals, exhibit the theory development proffered by Svalastog (see Chapters 1–5, this volume), demonstrating that the classic elements of location tied to physical resource have a clear resonance with tourists. Additionally, it is asserted that 'intangible' resources are sometimes the key to success for many tourist niche products. It also becomes apparent that particular festivals also exhibit an 'intangible' product, drawing many tourists to them, especially those within the segments of cultural and explorer tourism. In many respects, examining niche tourism products can be a useful lens to support Svalastog's (see Chapters 1–5, this volume) treatise. The following section uses a specific cultural product to evaluate niche development and Svalastog's theory.

The Case of Literary Festivals in the UK

Cultural tourism is seen as an important segment, representing a significant tourist typology in which there are a number of micro markets, festivals being one. Within the UK, there has been a growing importance of literary festivals to the tourist economy. A number of significant festivals and tourist destinations in the UK are associated with literature, in particular Edinburgh (Marques, 2019) and Hay (Jenkins, 2019). These annual festivals have important tourist cachet and demonstrate significant 'cultural capital' for those who attend (Bauman, 1999), clearly relating to a location's resources combined with intangible resource developments. Edinburgh and Hay literary festivals are also exemplars of the concept of a niche, both in terms of the 'structuralist and neoclassical paradigms'. Literary festivals are highly focused; their endeavour is fixed upon books and authors, demonstrating greater specificity than the segments of 'arts and events tourism', and, as identified in Svalastog's (see Chapters 1–5, this volume) work, the notion of intangible resources. Furthermore, the size of the market demonstrates a large variance in visitor numbers (5000–200,000). Literary festivals also offer a wide array of topics and foci, incorporating sales of newly published books, exploration of authors' works, celebration of particular authors, creative writing lessons and encouraging children's literacy skills. It is asserted that literary festivals are typical of the differentiated tourist product and divergent leisure experiences/products of a post-modern society. Therefore, literary festivals are useful case studies for exploring the value and importance of niche tourism products (Hoppen *et al.*, 2014) and their association with tourist destinations resources and product developments. Clearly, literary festivals have a wide appeal for the prospective visitor and tourist. Many of these festivals also seem to be linked with changes in consumerism over the last 50 years. This is evidenced by the longevity of some of the festivals, such as Cheltenham, which started in 1949. The British Council (2005) identifies 164 literary festivals, equivalent to nearly 30% of a UK festival total of 557, supporting the assertion that literary festivals are an important tourist resource.

Literary festivals also pose some important tourism questions, as many festivals clearly commodify literature, challenging the notion of authenticity and aspects of uniqueness (Jenkins and Lund, 2019). How authentic are these festivals or does this matter to the visitor and tourist? Authenticity can be viewed and defined in a number of ways, not simply through the idea that commodified 'objects' start to lose their authenticity. As Wang (1999) suggested, the post-modern perception of authenticity might lie with the 'tourist's experience' at the festival rather than the 'object' (author/books). Therefore, even the more commercialized festivals (those with large sponsorship from private companies) may still be seen as authentic by visitors. Certainly, there are protagonists who view high-profile literary festivals, such as Edinburgh and Hay, as lacking authenticity, but this is possibly in terms of the 'object' rather than the experience. Also, many of the more commercialized and commodified festivals offer the tourist/

visitor the opportunity of a 'genuine experience' in terms of access and presence of a book's author, rather than a virtual experience through television and other forms of media.

In view of the above debate, it seems pertinent to actually examine the definitions of literature and compare these to the aspect of commodification. The following are a number of definitions of literature (Wordnet, 2005, p. 1):

- creative writing of recognized artistic value;
- humanistic study of a body of literature: 'He took a course in Russian lit.';
- published writings in a particular style on a particular subject; the 'technical literature'; 'one aspect of Waterloo has not yet been treated in the literature'; and
- the profession or art of a writer: 'her place in literature is secure'.

In many senses, this consumed product can be viewed as a physical object (tangible), a book or text, containing stimuli (words, phrases and images) to evoke feelings and thoughts which are intangible. Also, books can become audible and enacted, such as in plays and literary recitals. Furthermore, a location or destination provides a cultural aura affording a sense of place to the performance. Hence, literary festivals appear linked to both tangible (books, location, tourism actors) and intangible resources of a destination (sense of place and evocation of the performances). In many cases they are perhaps 'transient' resources (intangible in Svalastog's treatise).

The Hay Literature Festival

The Hay Festival has all the elements of a niche tourism product as identified in the above discussion. These include: unique, authentic, specific resources being part of a neoclassical niche (see Fig. 9.1). The festival is based on two unique resource elements: first, the physical and geographical location of Hay-on-Wye, acclaimed to be the first second-hand-book town in the UK (possibly within Europe); and, secondly, the cultural component of the festival which is a cultural assemblage of authors and books designed to engage tourists at a temporary constructed festival site over a short period of time (referring to Svalastog's knowledge concept). The Hay Festival boasts international status but evidence suggests that it has a more nationally based tourist market, but is still considered extremely important in providing tourism in a rural destination. As stated by VisitWales:

> Hay Festival, one of the world's biggest and best literary festivals. For a town of 1500, Hay-on-Wye punches well above its weight. Famed as a book town ever since 'King' Richard Booth opened the first second-hand bookshop 50 years ago, the launch of Hay Festival in 1988 earned it a place on the literary and global map.
> (VisitWales, 2021)

It is considered to be an extremely important product for the destination in terms of economic income. Furthermore, it has expanded into a global literary market with a number of Hay festivals taking place around the world, a typical example being the festival held in Cartagena, Columbia (Procolumbia, 2021).

The Hay Festival is located in a small town close to the border of England and Wales (Meyers, 2002). The town has a population of some 1,300 people (1,500 in 2022) and its image is one of a diminutive urban location within the countryside, lying approximately mid-way between Brecon and Hereford (Meyers, 2002). It is located at a considerable distance from any major cities or towns, giving the destination image of Hay a countryside perspective and contrasting greatly from other UK major literary festivals, usually located in cities with excellent access routes and transport networks. Tourism in and around Hay is dominated by outdoor recreation and activities (PCC, 2005), encouraged by the close proximity of the Brecon Beacons National Park.

Hay lacks the Georgian architecture of Cheltenham or Bath and has little, if anything, in common with other festivals such as Swansea's, which is a large post-industrial city and port. Its main tourist resource appears to be the proliferation of second-hand-book shops (Florence, 2002), 39 in total (that in 2022 number about 15), which may be viewed as 'unique' within the UK. It is possibly this superstructure that gives gravitas to the Hay Literature Festival, creating its tourist destination image (Seaton, 1996) and supporting Svalastog's (see Chapters 1–5, this volume) treatise on resource allocation being essential to tourism product development. The emergence of Hay-on-Wye as a book town can be traced back to 1961 when Richard Booth pioneered this new form of economic development (Seaton, 1996). It is also asserted that the prominence of bookshops, coupled with the Hay Festival give this destination a competitive advantage over other similar rural towns within the area. Seaton would agree with this, noting that Hay-on-Wye was the first worldwide destination to be known as a 'book town' (Seaton, 1996). Book towns have now been established in Redu (Belgium), Becherel (Brittany) and Montouilieu (southern France) and there are now apparently some 17 book towns (Seaton, 1996). In 1995, it was estimated that Hay-on-Wye received 1,000,000 visitors each year, with most allegedly related to book tourism (Seaton, 1996).

Hay-on-Wye has been under the governance of Powys County Council (PCC) and in terms of tourism development, the Tourism Partnership Mid-Wales (TPMW) controls policies affecting the town. The tourism theme for the area, including Hay-on-Wye, has been that of 'naturally different', reflecting the rurality of the Tourism Partnership Mid-Wales spatial region. Although the Hay Festival was not directly mentioned within the tourism strategies of PCC and TPMD, festivals and events are recognized as important segments for Wales and also Mid-Wales (TPMW, 2005). Once again, this confirms Svalastog's (see Chapters 1–5, this volume) emphasis on the planning of tourist destinations based upon their available resources. They are identified under the segment of 'culture and cultural heritage', with events documented as being an important product for the region (TPMW, 2005), including the Hay Festival. The strategy notes:

> Events have been identified as an important draw for the region. On-going public sector financial support for events at all levels is declining and the need for a more sustainable approach has been recognised. TPMW will regionally lead on developing a more comprehensive approach to event support. This will build on the recent work being undertaken by Gwynedd Events Commission on behalf of TPMW and WDA.
>
> (TPMW, 2005, p. 13)

It also identifies 'short stay' visitors as being important, again relevant to the Hay Festival as tourists to the festival stay for only a few days.

The TPMW strategy also identified certain key tourism concepts such as 'authenticity' and 'distinctiveness' and the concept of sustainability. Seaton (1996) acknowledges that 'book tourism' offered by Hay-on-Wye is clearly in keeping with the notion of sustainability, inferring that the Hay Festival may be a part of this product. Furthermore, it supports Svalastog's (see Chapters 1–5, this volume) notion of location, using appropriate resources and creativity to provide a successful tourist product. Both Powys County Council and the Tourism Partnership Mid-Wales note the importance of tourism to this region as there are few other economic sectors that are successful within the area (PCC, 2005; TPMW, 2005).

Clearly, the Hay Festival has been firmly part of both the County Council's and the Regional Tourism Partnership's aim to encourage tourism to the area. Although not specifically identified as a niche, it might be asserted that this festival/event might be considered to be one, as the identified products within the strategy (TPMW, 2005), such as culture and cultural events, are stated to be segments. Therefore, specific festivals are part of the culture and cultural events segments and could be considered niches, based upon the resources of the location.

The Hay Literature Festival began in 1988 as a 'brand extension of the town's unique destination image of a "book town"' (Seaton, 1996, p. 280), supporting the theory of the need to have necessary resources present in a location to offer a successful tourist product. Since then, it has become one of the most successful literary festivals (*National Geographic*, 2021), rivalling that of Edinburgh and claiming to be of 'international' importance (Seaton, 1996; Florence, 2002). The eminence of the Hay Festival can be validated with events that include important celebrities; for example, the visit of ex-US president Bill Clinton (Jones, 2005). The Hay Festival of 2005 attracted such star names as Goldie Hawn, Jane Fonda and 'one of Hollywood's most influential film-makers', Spike Lee (Ezard, 2005; Jones, 2005, p. 6). Once again, the emphasis seems to be on post-modern commodities related to celebrity, rather than the authentic aspect of literature *per se*. It would seem that the success of the Hay Festival is contingent upon attracting top celebrities, more so than its focus on literature, even though the initial 'tourist product' for the festival was clearly based upon the unique selling point of a book town. However, shops sell books, so the commodification of a book festival is simply an extension of the 'culture capital' already found in Hay-on-Wye. Certainly, the 'cultural capital' of the festival can be considered to have a great propensity towards 'high art' products. This is supported by the festival sponsors who, compared with other literary festivals, demonstrate a more important role and greater control over the Hay Festival.

The opening pages of the previous festival programmes identify a number of important private companies. *The Guardian* was initially a 'title sponsor' but was replaced by *The Telegraph* (Hayfestival.org, 2013), signifying a likely tourist market of higher socio-economic class readers (Negrine, 1994; TGHF, 2005a). Sponsorship is divided up into specific commercial areas: Orange was the 'communication partner' (TGHF, 2005a) but was replaced by Sky Arts HD (Hayfestival.org, 2013). It also has many other sponsorship areas and as the festival has grown in attendance so have the number of sponsors. In 2003, there were only 15; by 2005 there were 31 UK national or international organizations, and in 2013, there were 50, suggesting a growing status and commercialization of the festival. The proliferation of commercial sponsors is also evidence of the prominence of the Hay Festival as a tourist product, as funding agencies such as the Welsh Development Agency (WDA), the Wales Tourist Board (WTB) and Powys County Council (PCC) were also prominent sponsors. The international status of the festival is confirmed by other sponsors such as AIA (Anglo-Israel Association), Canada House and the Israeli Embassy (TGHF, 2005a). Sponsors and sponsorship change from year to year; nonetheless, the elements of tourism and prominent commercial companies seem to be suffused throughout the festival, with each event sponsored by a commercial company. It is also interesting to note that the main sponsors are not necessarily publishing companies, as with the Bath and Cheltenham festivals, although Pembertons (publishing company) has been the single 'festival bookseller' (TGHF, 2005a) and literary company. An inference to be drawn from the wide range of sponsors at the festival is that the 'literature festival' is more than simply a book fair engendering a number of tourist and literary gazes.

It is also significant to note that the festival's name changed from the Hay Literature Festival to simply the Hay Festival. This name change seems in keeping with the programming of events and, as noted above, the prominence of celebrities who do not necessarily have connections with the destination, literature or books. Since 2000, there appears to have been a growing development of other forms of art such as music, dance and film. This move appears linked to the ever-growing number of tourists who attend the festival. In 1995, there were 25,000 visitors (Seaton, 1996); this has now risen to 80,000 (TGHF, 2005b). It is also interesting to note that the copyright in 2005 was Hay Festival of Literature and the Arts Ltd (TGHF, 2005b) and is now simply HayFestival.org (Hayfestival.org, 2013).

The programming for events has been similar to other literary festivals; the festival would run over a two-week period and incorporate ten days of events, starting at 9.00 am and concluding at 11.00 pm. This is a somewhat longer daily programme than other festivals, possibly reflecting the notion that visitors are actually staying in or around Hay and wish to see a number of events during the day. In 2005, there were 359 events over ten days, compared to 249 in 2003 (TGHF, 2003; 2005a), but in 2013 this was down to 200 (Hayfestival.org, 2013). The number of film and other art events are significantly minor to the main literary events. The festival weeks are always the last week of May and the first week of June.

The Hay Festival uses a number of different venues, many of which are temporary, such as marquees. The central festival site has changed from the town's primary school to agricultural fields adjacent to the town. There are numerous temporary structures used for the festival, hosting different events, and many of these tents/marquees symbolize specific festival sponsors, emphasizing the commodification of the festival (TGHF, 2003; 2005a). The media coverage of the Hay Festival is decidedly prominent, with daily reports printed in the title sponsor's newspaper and BBC websites, with many of these updates being covered by other media outlets.

Summary and Inferences

It is obvious that today's tourism is different to that of the mid-20th century. Changes in tourism reflect the changes in market economies, which appear linked to the concept of modernity emerging into post-modernity. This is best summarized by Urry (2002) who identifies economic markets as having two elements. First, the products of mass production can still be seen within the tourism package, which continues to dominate the tourism market. For example, Spain still supports its original markets based upon 'sun, sand and sangria' (Sabau, 2005). Secondly, there has been a marked shift towards diversity of products and the importance placed upon the consumer rather than the producer. This can be viewed as a demassification and splintering of traditional markets towards niche markets.

The issue of globalization is also part of structural economic change. For tourism, the compression of space and time is a critical element in the composition of the tourism product, affecting the markets and products offered. These two features, change in consumption (Hoppen *et al.*, 2014; Sert, 2017) and globalization, have engendered the emergence of niche tourism markets and products (Robinson, 2012). The use of the word 'niche' seems to have universal approval and understanding; certainly, many NTOs did cite it regularly. The tourism products on offer illustrate a move to more specialized markets supporting the viability of niche products and markets. Niches are closely connected with segmentation, and segmentation is, in essence, the *raison d'être* of marketing (Statt, 1997).

Clearly, texts and examples of perceived niches suggest that a singular definition is not acceptable. The discussion on niches identifies that other elements may be applied to niches, such as service and expectations (Robinson and Novelli, 2005). Some authors are now recognizing that an economic niche has other meanings and representations which are non-economic, supporting the dictum of Baudrillard's (1999) semiotic purchase of goods, whereby goods and services supply meaning and satisfaction to the consumer. This also corroborates Svalastog's (see Chapters 1–5, this volume) assertion of intangible resources being important to a destination's success.

Robinson and Novelli (2005) support these assertions and recognize that niches have a number of meanings. Furthermore, imbued within all niches are aspects of 'The connotation of a more tailored and individualised service [carrying] its own cachet relating to the features like small scale of operations, implied care and selectivity regarding discerning markets and a suggested sensitivity to tourist' (Robinson and Novelli, 2005, p. 6). This, once again, supports Svalastog's theory that tourist destinations are based on resources at a destination and it appears that many resources are now being used in a new way to offer fresh experiences to the tourist, the Hay Festival being one example of this. Geography is still key in many niche tourist products and, as illustrated with this case study, Hay's development is clearly linked to the unique feature of the destination (book town). This, firstly, relates to the built physical fabric of the destination known for second-hand books (Jenkins, 2019). From this emerged, via an intangible resource (the idea of a book festival), the establishment of the Hay Festival. Although based on the theme of literature, it has been elevated to allow experiential tourist experiences by providing access to authors and opportunities to listen to their verbal narratives.

It can be shown from the text on the case study that the Hay Festival is a niche market following the classic niche structure. It is clearly unique, given its location, in both the UK and perhaps Europe, offering a plethora of second-hand-book shops (physical resources). The evidence suggests that it is the experiential element of the festival that is essential to its success (intangible resource). The physical supply of tourist infrastructure is extremely limited with only a few small hotels and B&Bs, and these are often booked a year in advance. Even with the limited tourist infrastructure, the festival attracts large numbers of visitors (80,000 or more) who stay in the region for a few days and possibly contribute up to six months worth of GDP to the town (Jenkins, 2019).

Niche products/destinations envelop a number of cultural and social areas and are clearly enhanced by a multidisciplinary approach to their development. Economic factors, psychological motivations and social environments all contribute to a festival's success, including aspects of the cultural heritage and tenacity of a destination. These factors are frequently overlooked and ignored by tourism planners.

Hay's success is clearly based on the localization of resources, which are marketed to international, national and regional audiences

(VisitWales, 2021). Niches have now become a mainstay of many tourist products and strategic plans even though there are misunderstandings between what exactly defines a niche. It is evident that the concepts of niche products and markets have a number of meanings and interpretations but the overall inference is manifestly ascribed as the 'antithesis of mass tourism'.

References

Adventure Associates (2005) Scott & Shackelton's Antarctica, Christchurch to Hobart – aboard MV Orion. Available at: http://www.adventureassociates.com/antartica/ross/scottshackelton (accessed 22 June 2006). Adventure Associates, Sydney, Australia.

Baudrillard, J. (1999) *The Consumer Society, Myths and Structures*. Sage, London.

Bauman, Z. (1999) *Culture as Praxis*. Sage, London.

British Council (2005) *Literature Festivals*. British Council, London.

BTA (2002) Sports tourism in the UK. Research report based on IPS data. BTA (British Tourist Authority), London.

Choat, I. (2004) Big operators eye niche markets. *Guardian Unlimited*, 4 Dec. 2004. Available at: www.travelguardian.co.uk/saturdaysections/story/0.8922.13658060 (accessed 1 July 2005).

Dahringer, L.D. and Muchlbacher, H. (1991) *International Marketing*. Addison-Wesley, Wokingham, UK.

DCMS (1999) *Tomorrow's Tourism: A Growth Industry for the New Millennium*. Department for Culture, Media and Sport (DCMS), London.

DCMS (2003) *Strategic Plan 2003–2006*. Department for Culture, Media and Sport (DCMS), London.

DCMS (2011) *Government Tourism Strategy*. Department for Culture, Media and Sport (DCMS), London.

Doyle, B. (2003) The holy grail of film tourism – or how to increase visitor numbers without really trying. In: Macleod, D.V.L. (ed.) *Niche Tourism in Question*. Conference Proceedings, University of Glasgow, June 2002. Critchton Publications, Dumfries, UK, pp. 101–104.

Ezard, J. (2005) Guardian Hay Festival turns to film. *The Guardian*, 17 February.

Florence, P. (2002) Hay Festival website: http://www.hayfestival.co.uk/site/index.htm (accessed 29 January 2003).

Hayfestival.org (2013) *The Times* Hay Festival 2013. Hayfestival.org.

Hoppen, A., Brown, L. and Fyall, A. (2014) Literary tourism: opportunities and challenges for the marketing and branding of destinations? *Journal of Destination Marketing & Management* 3(1), 37–47.

Jenkins, I.S. (2007) Postmodern tourist niches: the case of literary festivals. PhD theseis, University of Wales, Swansea, UK.

Jenkins, I.S. (2019) The Hay Festival: a longitudinal study on its attributes and the sustainable impacts on a small Welsh town. In: Jenkins, I.S. and Lund, K.A. (eds) *Literary Tourism: Theory, Practice and Case Studies*. CAB International, Wallingford, UK, pp. 120–132.

Jenkins, I.S. and Lund, K.A. (eds) (2019) *Literary Tourism: Theory, Practice and Case Studies*. CAB International, Wallingford, UK.

Jones H. (2005) Film theme will add Hollywood glamour to Hay festival line-up. *Western Mail*, 18 February.

Kotler, P. (2000) *Marketing Management: The Millennium Edition*. Prentice-Hall International, London.

MacLellan, R. (2003) Niche tourism markets in Scottish tourism: an investigation of niche markets reliant on the natural environment. In: Macleod, D.V.L. (ed.) *Niche Tourism in Question*. Conference Proceedings, University of Glasgow, June 2002. Critchton Publications, Dumfries, UK.

Marques, L. (2019) The making of the literary city: Edinburgh, Barcelona, Óbidos. In: Jenkins, I.S. and Lund, K.A. (eds) *Literary Tourism: Theory, Practice and Case Studies*. CAB International, Wallingford, UK, pp. 57–70.

Marson, D. (2011) From mass tourism to niche tourism. In: Robinson, P., Heitmann, S. and Dieke, P.U.C. (eds) *Research Themes for Tourism*. CAB International, Wallingford, UK, pp. 1–11.

Meyers C.T. (2002) Three sights to savour. Available at: http://www.biztravel.com/ARTICLES/travel3.html (accessed 29 January 2003). Infohub, California.

National Geographic (2021) The travel destinations that inspired this summer's hottest novels travel. *National Geographic*, 26 May 2021. Available at: https://www.nationalgeographic.co.uk/travel/2021/05/hay-festival-writers-travel-inspiration (accessed 18 April 2022).

Negrine, R. (1994) *Politics and the Mass Media*. 2nd edn. Routledge, London.

O'Reilly A. (1999) Special reports 25/1/99 – Bord Fáilte focus on niche markets in 1999. Adworld, Bell Rock Communications, Co. Wicklow, Ireland. Available at: www.adworld.ie/special_reports.htm (accessed 10 June 2001).
Page, S.J., Bunt, P., Busby, G. and Connell, J. (2001) *Tourism: A Modern Synthesis*. Thomson Learning, London.
Papathanassis, A. (2011) *The Long Tail of Tourism: Holiday Niches and Their Impact on Mainstream Tourism*. Gabler, Wiesbaden, Germany.
PCC (Powys County Council) (2005) *A Sustainable Tourism Strategy for Powys: Making Quality Count*. PCC, Llandridnod Wells, Wales.
Pender, L. and Sharpley, L. (eds) (2005) *The Management of Tourism*. Sage, London.
Procolumbia (2021) Literary cartagena: Hay Festival. Available at: https://colombia.travel/en/blog/literary-cartagena-hay-festival (accessed 29 September 2021).
Ritzer, G. (1996) *The McDonaldization of Society*. Pine Forge Press, London.
Robinson, M. and Novelli, M. (2005) Niche tourism: an introduction. In: Novelli, M. (ed.) *Niche Tourism: Contemporary Issues, Trends and Cases*. Elsevier, London.
Robinson, P. (ed.) (2012) *Tourism: The Key Concepts*. Routledge, Abingdon, UK.
Rossetti, G. and Quinn, B. (2019) Learning at literary festivals. In: Jenkins, I.S. and Lund, K.A. (eds) *Literary Tourism: Theory, Practice and Case Studies*. CAB International, Wallingford, UK, pp. 93–111.
Sabau, M.F. (2005) More than just fun in the sun? *Locum Destination Review*. Available at: www.locum-destination.com/pdf/LDR9MoreThanJustFun.pdf (accessed 23 June 2005).
Seaton, A.V. (1996) Hay-on-Wye, the mouse that roared: book towns and rural tourism. *Tourism Management* 17(5), 279–285.
Sert, A.N. (2017) Niche marketing and tourism. *Journal of Business Management and Economic Research* 1, 14–25.
Shaw, G. and Williams, A.M. (2002) *Critical Issues in Tourism: A Geographical Perspective*. 2nd edn. Blackwell, Oxford, UK.
Statt, D.A. (1997) *Understanding the Consumer: A Psychological Approach*. Macmillan Press, London.
STB (Scottish Tourist Board) (2005) VisitScotland, Literary Scotland. Available at: www.visitscotland.com/sitwide/fivestarfeatures/literaryscotland (accessed 13 July 2005).
Scottish Parliament (2002) Niche tourism in Scotland: golf and business tourism. SPICe briefing, Enterprise and Longlife Learning Committee, 02/94, 21/8/2002. The Scottish Parliament, Edinburgh.
TGHF (The Guardian Hay Festival) (2002) *Change your mind. The Festival Guide*. Hay-on-Wye Festival.
TGHF (The Guardian Hay Festival) (2005a) *Tilting at Windmills*. 26 May–5 June.
TGHF (The Guardian Hay Festival) (2005b) The Guardian Hay Festival, 2006. Homepage. 26 May–5 June.
Tourism Australia (2005) Niche experiences. Available at: www.tourism.australia.com/Markets.asp?lang=EN&sub=0338 (accessed 21 June 2005).
TPMW (Tourism Partnership MidWales) (2005) Business plan for Tourism Partnership MidWales, 2005/6. Available at: http://www.tpmw.co.uk/files/tpmw_busplan_05-06_eng.pdf (accessed 6 July 2005).
Urry, J. (2002) *The Tourist Gaze*. Sage, London.
VisitWales (2021) Introducing the Hay Festival. Available at: https://www.visitwales.com/things-do/events/festivals/introducing-hay (accessed 29 September 2021).
Wang N. (1999) Rethinking authenticity in tourism experience. *Annals of Tourism Research* 26(2), 349–370.
Wordnet (2005) Definition of literature. Available at: http://wordnet.princeton.edu/perl/webwn?s=literature (accessed 30 December 2005). Cognitive Science Laboratory, Princeton University, Princeton, New Jersey.
WTB (2002) A proposed strategy for golf tourism in Wales. Wales Tourist Board, Cardiff.
WTB (2004) *Autumn Glory: Walking Wales* (marketing leaflet). Wales Tourist Board, Cardiff.

10 Living Environment and Attractiveness

Lars Aronsson

Perspectives and Research Objectives

Sweden ought to be the best country in which to reside, work and grow. This will be achieved by providing people, businesses, municipalities and regions with the preconditions to create attractive environments and shape their future based on their unique circumstances.

This translated quote can be found on the Ministry of Enterprise and Innovation's programme platform 'Attraction Sweden' (*Attraktionskraft Sverige*, Nylen, 2012), which was intended to support work on issues and questions central to local and regional attractiveness. The quote demonstrates that attractiveness and quality of life are featured on the political agenda. The present study focuses on these issues in the region of Värmland, but the reasoning and analysis that is made is also valid for other places and regions with similar structural features and should therefore be of general interest.[1]

An appealing and attractive living environment has two elements; on the one hand is a question of *insider perspective* – how residents of a place or region view their environment – and, on the other hand, it is also a question of *outside perspective* – what image is communicated and perceived from outside.

This means that it is of great interest to gain an impression of how people in different living environments view their immediate surroundings in terms of both the natural conditions as well as social and service conditions. It is also of interest to gain an idea of the images of Värmland and how the region is perceived externally. From this perspective the living environment becomes an important element integrated into discussions on the attractiveness of the region. This is important, not least for tourism. Attractive environments are thus important for attracting visitors and in-migrants; people are expected to choose to live in a place that they find attractive and attractiveness is also expected to form a location factor for investment and business.

The purpose of this study is to explore the attractiveness of the living environment from both the insider and outsider perspectives. This is done using a meta-analysis of other studies on living environments and attractiveness in Värmland and, although based on secondary data, is unique as it combines perspectives and findings from different studies forming one overall picture.

The idea and experience of attractiveness in a living environment is multi-dimensional and complex. It can be experienced as attractive to varying degrees and in different forms, depending on circumstances and contexts. Its perception also varies according to time of life and personal experiences. There may be a longing for both the home environment and 'other' places that have been visited and experienced and this study

DOI: 10.1079/9781800621480.0010

attempts to understand perceptions of the living environment in Värmland in different ways and from different perspectives.

Points of Departure

A point of departure for the study is the difference between what can be called insider and outsider perspectives. Insider perspectives concern how Värmlanders view and assess their living environment – defined as a residential area, municipality or county – in Värmland. Surveys reflecting an outsider perspective are often focused on national images of Värmland but also around images of Värmland generated by students and families from large cities. In an attempt to describe an outsider's perspective, the chapter also considers some studies of Värmland by the media.

The study also covers perceptions of residents, businesses and visitors, often challenging insiders' views of the images and narratives presented of Värmland. Regional development, public planning and tourism development address both general development traits and specific conditions.

The constructed images of insiders, i.e. the perceptions of the population of Värmland regarding their living environment, are analysed based on different studies undertaken after 2005.

The studies used in this chapter are:

- people of Värmland, their lives and health (Kalander Blomqvist *et al.*, 2008). The author's analysis is based on the empirical material of this study;
- a survey conducted among the region's citizens in 2010 and early 2011, published in the book *Värmländska Landskap* (*Landscapes of Värmland*, Nilsson *et al.*, 2012);
- Statistics Sweden (SCB) citizen surveys in various municipalities in Sweden (Arvika, 2011; Eda, 2011; Hagfors, 2008; Hammaro, 2011; Karlstad, 2010; Kil, 2010; Kristinehamn, 2010; Storfors, 2011; Sunne, 2010; Saffle, 2011; Torsby, 2011).[2]

The outsider perspective describes different images of Värmland found in studies of media images, important decision makers and students and families in big cities. Furthermore, consideration was given to empirical material examining Värmland's brand value. The latter studies covered Värmland as a brand, perceptions of the regional business community, Värmland as a residential alternative and its attractiveness (Attityd, 2001, 2004, 2008, 2009, 2010; Observer i Sverige, 2005; Värmland i media, oktober 2004–mars 2005; Observer i Sverige, 2006; Värmland i media, oktober 2005–mars 2006; Cision Sverige, 2007; Värmland i media, oktober 2006–mars 2007; Infopag Sverige, 2010; Värmland i media, oktober 2009–mars 2010; Gromark and Anselmsson, 2011; Cision Sverige, 2007).

Värmland's Living Environment and Attractiveness

As a consequence of using secondary data, the study offers many different perspectives on the living environment and attractiveness of the location. This poses a challenge in that it is difficult to select and summarize aspects of the data. Another challenge is that concepts such as 'living environment' and 'attractiveness' are complex and contain many dimensions. The study identified differences among different generations and between different geographical areas of the county. The population in Värmland recognizes there are challenges; for example, when it comes to employment. At the same time they generally see Värmland as a positive living space. There is a dissonance between residents' and non-residents' views of Värmland relating to knowledge of it within the region. Although the study focuses on Värmland, the whole region shares Värmland's problems, including the impacts of urbanization and depopulation, reflected in out-migration of inhabitants and business changes.

This study is significant due to the ongoing discussion on the future of Värmland manifested, for example, in efforts to foster new regional development plans. The idea of attractiveness of places and regions is becoming increasingly important in the context of regional development and regional growth policy. The discussion, however, suggests that the issues of good living environments and attractiveness are elusive.

The attractiveness of places and regions and their living environments can be studied using different scientific approaches (see, for example, Relph, 1976; Tuan, 1977; Paasi, 2003; Massey, 2005; Cresswell, 2013). The perspective and focus for this study is human geography. Theoretical starting points include issues of regional development and the fact that not all regions and locations have the same preconditions and thus do not evolve in the same way. Besides regional development, applied theories briefly address the importance of culture for locales and for marketing a sense of place. Hence, all theoretical discourses in this chapter have included ideas on the role of good living environments and on attractiveness of places.

Places and cities of differing size contain, and are shaped by, various 'lives', such as business life, cultural life, social life and more. This implies that there are many stakeholders in social development at the locale. An important factor in a person's choice of residence is the perception of good living environments. In this context, all aspects of life are considered including opportunities for work, education, access to public and private services, local culture and leisure opportunities, social contacts and access to healthcare. It can be postulated that good housing may be seen in the context of social interaction and the attractiveness of the surrounding built or natural environment.

Culture, often linked to the development of urban areas, has been prominent in Sweden for several decades (Andersson, 1985) and can be found in both policy and research issues. However, Richard Florida (2002) has renewed an interest in culture and helped to reinstate it on the agenda. It is linked to local and regional development, primarily because Florida's work had a major impact on practice and because his theories have inspired urban development strategies all over Sweden. In contrast, Florida's ideas have been widely debated and criticized in scientific circles (Aronsson and Braunerhielm, 2011).

Today, practitioners of urban planning are engaged in the challenge of how to integrate the physical environment and architecture in urban planning with cultural expressions, symbols and aestheticism. It aims to attract more residents, encouraging people to choose to reside in a particular locale. Furthermore, it seeks to attract visitors and investment resulting in local and regional economic growth. Therefore, professionals in planning and place marketing create emotive frames, such as branding of place and imaging attractive physical environments (Aronsson *et al.*, 2007). It is considered important to develop an interesting image or narrative about a place that will attract people and will be based on innovative housing schemes and tourism businesses, utilizing both the urban and natural landscapes of the locale. The developments will be constructed upon proximity to water, recreational areas, architecture, cultural heritage, pedestrian areas, conservatories etc., including an attractive range of commercial spaces (such as malls or shopping centres). Moreover, temporary venues for creative encounters such as festivals, conferences or activities in culture and sports should also enhance the attractiveness of place. In summary, it is considered crucial to develop place-specific attributes based on diversity, vibrant cultural life, local identity, global reach and openness, competence, networks, meeting places and attractive environments. It would be expected that places offering these attributes can be considered to be attractive living environments for residents and also for attracting creative people. These ideas are primarily based on (large) city developments and are not commonly found in less populated urban and rural places.

The above discussion shows a growing interest in 'alternative' community planning methods. For example, cultural planning has become topical in the development of place (Ghilardi, 2003; Lundberg and Hjort, 2010).

The study demonstrates that the living environment is a complex concept encompassing many dimensions of living and working in a place or region. Living environments are multi-faceted and include housing and how employment, education, services, culture and leisure opportunities, and more, are accessed. But it is also about business development and the willingness of businesses and industry to nurture and invest in place. The attractiveness and appeal of a place or region are also reflected in the development of leisure opportunities, including the tourism industry and the cultural and creative industries (cf. Svalastog, 1994, 1998).

Living environments have different geographies and can encompass everything from residential areas and the municipality to the county and region. The living environment and the attractiveness of regions and places can be experienced in different ways. Familiarity and experiences play a role in how we look at and assess living environments in different places. The more superficial the former, the more stereotypical is the image. But even from a local's perspective, the image may vary due to social status, life cycle and residential location within the microgeography of the locale. Presumably, these differences between generations are often generic and not specific to Värmland, but this does not make the differences irrelevant. The gap between generations reveals a 'mental urbanization' in that many young people's values and attitudes are positive about the places and regions that provide opportunities for education, work, career and housing, which is often associated with urban areas.

Värmland, in many ways, shares its situation and regional problems with other counties and regions in Sweden as well as across Europe. Urbanization is a powerful process and one that has gained new impetus in Sweden since the early 2000s. Urbanization is not heralded by all and there are factions within the planning community that challenge the ideals and values of this process.

Results and Reflections: An Insider Perspective

Broadly speaking, the inhabitants of Värmland appear happy with their residential area. A majority believe that it is a safe and secure region and that the environment is good for bringing up children. Furthermore, social contacts are generally considered good with a community based on mutual trust. The inhabitants of Värmland seem to like where they live, regardless of whether this is a municipal centre, a small town or the countryside. However, there are clear differences between municipalities. According to Statistics Sweden's surveys, the most satisfied are the inhabitants of Karlstad, Hammarö and Sunne, while the least satisfied are those living in Storfors, Eda and Hagfors. As has been shown in this discourse, it is difficult to standardize the components of a good living environment. As stated previously, the living environment is complex and multi-dimensional, consisting of many parts, but it is also holistic. We believe it is important to try to find out how people experience the living environment in different parts of the county, which information can then be used in discussions on the future development of Värmland.

In the eyes of the population in Värmland, nature is a more important element for a good living environment than culture. (Culture, in this context refers to aspects of fine art.) Service issues are also important and services could be nearer or more available than they are currently. Closure of services and their centralization are seen as increasingly problematic in many places. Even if residents become accustomed to it, the erosion of service provision plays a significant role in affecting the value of the living environment in the long term.

The differences between men and women and their assessments of the living environment can be considered minor. However, age is a factor affecting residents' perspectives on their environment. Older people often have a more positive attitude to what is available, compared with younger people. This is a fact probably shared with many other counties.

Various surveys show large similarities in results and outcomes, illustrating a great stability over time in residents' perceptions of the living environment and of the county's attractiveness.

Results and Reflections: An Outsider Perspective

Residents believe that other locals hold similar positive views of images and values regarding the locality, especially those born in Värmland. This belief also extends to those living outside the immediate area of Värmland.

Studies based upon media images of Värmland and studies from the 2000s essentially highlight a positive regional business community. It is competitive, export-oriented and invests locally in Värmland. Even cultural life and, to some extent, the tourism industry is portrayed as a regional stronghold with development potential. Questions concerning housing and living environments have no major role when it comes to media coverage of Värmland. Regional policy documents highlight something genuine and unique and it can be concluded from the latest media survey that Värmland is presented and experienced as something remote and off the beaten track. However, this is not necessarily a positive image.

Place branding is a part of regional development and for decades there have been clear intentions and initiatives to increase awareness of Värmland and to advertise or sell it in various ways. As part of the monitoring and evaluation of these regional activities, a number of market surveys were conducted about how people – often those in power in business, politics and organization – envisage Värmland. The investigations focused on non-resident perspectives of the county and the results show that the image of Värmland has not really changed.

The imagined view of Värmland differs between residents and visitors. Local residents' image is often linked to cultural identity, while visitors see a commercial image which is consistent with the media image, i.e. the image deliberately communicated to visitors (Wahlstrom, 1984, 1994). It can be concluded that it is difficult to change the image of a region or city in just a few years; it will probably take a much longer time to change the perception of Värmland among visitors. It is important to consider that the representative value and the image identified by the respondents who participated in the studies are strongly linked to the tourism industry in Värmland. Värmland's branding tries to create an image for visitors that includes culture, nature, sports, 'great and happy Värmlanders', dialects, stories, motorsport rallies, moose and so on. The strategy fails to address the wider business environment and is not targeted towards attracting other entrepreneurs or new residents to the location. A conclusion must therefore be that the current branding strategy does not address wider community needs. It is therefore essential to create a new image which will attract new businesses and residents to the area and to strengthen the image of entrepreneurship and business within Värmland. Knowledge of Värmland's potential economic development is poor outside the county itself and is perhaps more negative than that of the residents.

There are a surprising number of respondents who know nothing at all about Värmland, who barely know where the county is and who cannot identify images of it. Värmland does not seem to have a clear 'spot on the map.' To communicate unique properties outside the region's borders is all about place marketing. A region with a strong profile and a strong image has major advantages in terms of attracting visitors, businesses and new residents (Morgan *et al.*, 2002). It also has major advantages when it comes to maintaining population numbers in a location, together with strengthening residents' self-esteem and encouraging local businesses to take risks. Clearly, place marketing is about highlighting local and regional characteristics and unique characteristics, in the form of culture, nature and consumption (O'Dell, 2002), which is currently directed towards tourism rather than other business segments. For example, Löfgren (2005, p. 33) writes the following about this relationship: 'All over the world we encounter similar projects, which try to create a local profile using a global, cultural image. The risk is to create repeat imitations and trivialization of a product that cannot take advantage of a place's own special conditions' (author's translation).

Hospers (2010) argues that place marketing is far more complex than product marketing. Places influence people in a completely different way than products. They have meaning, often emotional in the form of place attachment and through socio-economic ties. According to Hospers, most cities use what he terms 'cold marketing', which briefly means transferring product marketing's traditional ideas to place marketing. Hospers points out that it is mandatory to separate target groups such as residents, in-migrants, tourists and businesses. For residents, in-migrants and, to some extent, businesses, Hospers advocates so-called 'hot marketing', which is directed towards people who have emotional or socio-economic ties to the current place. Hospers notes that this is equivalent to a spatial relationship in marketing (cf. Hankinson, 2004). The overarching question therefore is: How can existing target groups be retained in a location? Hospers argues a new approach needs to be considered, based upon those already living in the place, their needs and their perceived place values, all of which contribute to a good and attractive living environment. Another important group can be those who have moved from the 'place' but maintain emotional ties and who could, in the future, be attracted back to the area. To some extent, the same reasoning can be applied to businesses. However, for tourists and some new residents, the tourist image seems to be more important, meaning that these groups are affected by more traditional product marketing.

Changing the image of a place is a long-term process and may require a number of messages related to different target groups that one is trying to attract. There will then be different messages and images and it is important for Värmland to reflect on who is considered the most important target groups.

Women currently seem to be less affected by the image creation message of Värmland which can be viewed as both positive and negative. Creation of a new more female oriented message may well help in changing this, demonstrating that the old messages have not been effective. This will also provide new opportunities to attract this group who currently have little knowledge or image of the region.

However, there is a need to reflect on why perceptions between residents and visitors differ on the regional business community and regional brand. If local entrepreneurs are able to send a positive message to business outside of the location then it would be hoped that more new businesses could be attracted to the area. Currently, studies show that there are relatively few who are prepared to live, work or establish a business in Värmland.

In order to attract both new residents and business, there is a need to demonstrate the current success of the business environment, highlighting good employment opportunities and an attractive business environment. This is as important as a positive living environment and can be as crucial in attracting new residents as any marketed image. A city with an attractive built environment and vibrant cultural life that attracts employees is also a prerequisite for the establishment of companies (cf. Stockholms Handelskammare, 2008). The cultural fabric of Värmland is identified as an important feature, mirrored in the context of folklore, happy locals, music, stories and events. However, this is a culture mostly catering for visitors but which could also attract further activities for both the residents and the business community.

Conclusion

A conclusion that can be made is that the characteristic traits of Värmland are also important images of the region. These images are strongly associated with the place and region and may vary depending on whether they are linked to residents or visitors and what relationship visitors have to the place. The place can be seen as attractive as it offers opportunities for employment and various types of business development. This sends positive signals that are important when it comes to creating an attractive living environment. These signals also contribute to a stronger self-esteem of residents, implying that local businesses can risk new ventures such as startups which help to retain young people and stop depopulation. Furthermore, it might also attract back locals who have moved away. A positive living environment may also attract further government initiatives.

Every location attempts to sell its image through marketing, which has many channels and materials to utilize. The main aim of these will be to attract new businesses and visitors through ventures such as sponsorship, trade fairs, websites, social media, printed materials and various other kinds of advertisements. The effectiveness of each method needs to be reviewed but the uniqueness of place must reflect residents' perceptions while also attempting to attract new business and visitors, some of whom may eventually choose to reside in areas such as Värmland.

Note

1 The chapter is based on the following working paper: Aronsson, L., Braunerhielm, L., Dahlgren, B. and Westlindh, S. (2013) Livsmiljö och attraktivitet. Upplevelser inifrån och bilder utifrån av Värmland. Karlstad University Press, Karlstad, Sweden.

2 Listed here are the municipalities and dates of the surveys.

References

Andersson, Å.E. (1985) *Kreativitet storstadens framtid: En bok om Stockholm*. Bokförlaget Prisma/Regionplaneförbundet, Stockholm.

Aronsson, L. and Braunerhielm, L. (eds) (2011) *Kreativitet på plats*. Karlstad University Press, Karlstad, Sweden.

Aronsson, L., Bjälesjö, J. and Johansson, S. (eds) (2007) *Kulturell ekonomi: Skapandet av värden, platser och identiteter i upplevelsesamhället*. Studentlitteratur, Lund, Sweden.

Aronsson, L., Braunerhielm, L., Dahlgren, B. and Westlindh, S. (2013) *Livsmiljö och attraktivitet: Upplevelser inifrån och bilder utifrån av Värmland*. Karlstad University Press, Karlstad, Sweden.

Attityd (2001) Värmland låter spännande, våren och hösten 2001. Attityd i Karlstad AB, Karlstad, Sweden.

Attityd (2004) Uppfattningar om Värmland – nollmätning. Attityd i Karlstad AB, Karlstad, Sweden.

Attityd (2008) Är Värmland mer spännande – eftermätning. Attityd i Karlstad AB, Karlstad, Sweden.

Attityd (2009) Karlstad som studentstad. Attityd i Karlstad AB, Karlstad, Sweden.

Attityd (2010) En trevligare upplevelse – kännedom och attityder om Värmland 2009–2010. Attityd i Karlstad AB, Karlstad, Sweden.

Blomqvist, M.K., Janson, S. and Starrin, B. (eds) (2009) *Värmlänningarnas Liv och Hälsa 2008*. Karlstad University Press, Karlstad, Sweden.

Cision Sverige (2007) Värmland i media, oktober 2006–mars 2007. Cision Sverige, Stockholm.

Cresswell, T. (2013) *Place: A Short Introduction*. John Wiley & Sons, Chichester, UK.

Florida, R. (2002) *The Rise of the Creative Class...and How It's Transforming Work, Leisure, Community, and Everyday Life*. Basic Books, New York.

Ghilardi, L. (2003) Cultural planning: an integrated approach to cultural development. *Riksantikvarieämbetet 2003:7: Plats, drivkraft, samhällsprocess. Vad gör kulturarvet till en resurs för hållbar regional utveckling? Redovisning från seminariedagar i Uppsala 22–23 oktober*. RAÄ, Stockholm.

Gromark, J. and Anselmsson, J. (2011) Varumärksmätning Västra Götaland. Label Varumärkesutveckling, Göteborg, Sweden.

Hankinson, G. (2004) Relational network brands: towards a conceptual model of place brands. *Journal of Vacation Marketing* 10, 109–121.

Hospers, G.J. (2010) Making sense of place: from cold to warm city marketing. *Journal of Place Management and Development* 3, 182–192.

Infopag Sverige (2010) Värmland i medierna, oktober 2009–mars 2010. Infopag, Stockholm.

Löfgren, O. (2005) Upplevelseindustrin ska få landet att leva upp. *Axess* 4(4), 31–34.

Lundberg, K. and Hjort, C. (2010) *Att fånga platsens själ. Handbok om cultural planning*. Sveriges Kommuner och Landsting. SKL Kommentus Media AB. Stockholm.

Massey, D. (2005) *For Space*. Sage, London.

Morgan, N., Pritchard, A. and Pride, R. (eds) (2002) *Destination Branding: Creating the Unique Destination Proposition*. Butterworth Heinemann, Oxford, UK.

Nilsson, L., Aronsson, L. and Norell, P.O. (eds) (2012) *Värmländska landskap: Politik, ekonomi, samhälle, kultur, medier*. Karlstad University Press, Karlstad, Sweden.

Nylén, C. (2012) *Attraktionskraft Sverige*. Enheten för regional tillväxt, Näringsdepartementet, Regeringskansliet, Stockholm.

Observer i Sverige (2005) Värmland i media oktober 2004–mars 2005. Observer, Stockholm.

Observer i Sverige (2006) Värmland i media oktober 2005–mars 2006. Observer, Stockholm.

O´Dell, T. (ed.) (2002) *Upplevelsens materialitet*. Studentlitteratur, Lund, Sweden.

Paasi, A. (2003) Region and place: regional identity in question. *Progress in Human Geography* 27, 475–485.

Relph, E. (1976) *Place and Placelessness*. Pion, London.

Statistics Sweden (n.d.) SCB:s medborgarundersökningar. Available at: www.scb.se/medborgarundersokningen (accessed 18 April 2022).

Stockholms Handelskammare (2008) *Kreativitet och konkurrenskraft – en kritisk granskning av rådande teorier om urban utveckling*. Rapport 2008:2, Stockholms handelskammare, Stockholm.

Svalastog, S. (1994) *Lokalisering av reiseliv: Om ressursanalyser, den romlige fordeling og lokal innpassing*. 2. utgave. Högskolen i Lillehammer, Avdelning for planlegging, ledelse og reiselivsfag, Lillehammer, Norway.

Svalastog, S. (1998) *The Nature of Resources for Tourism and Recreation*. Högskolen i Lillehammer, Avdelning for planlegging, ledelse og reiselivsfag, Lillehammer, Norway.
Tuan, Y.-F. (1977) *Space and Place: The Perspective of Experience*. University of Minnesota Press, Minneapolis, Minnesota.
Wahlström, L. (1984) *Geografiutveckling och geografisk utveckling – som om platser betydde något...* Meddelanden från Göteborgs Universitets Geografiska Institutioner, Serie B, no. 76, Göteborg, Sweden.
Wahlström, L. (1994) *Bilder av platser – ett kulturgeografiskt perspektiv*. Choros 1994:4. Kulturgeografiska institutionen. Göteborgs Universitet, Sweden.

11 Heritage Trails through Dolenjska and Bela krajina in Slovenia: Tourism Entrepreneurship in Action and Stakeholders' Relationship

Marko Koščak

Overview

The Heritage Trail of Dolenjska and Bela krajina case study has a rural base and is profoundly affected by the requirement to attract tourism without damaging the sensitivities of the rural environment. There are several reasons why this case study was chosen. First, to present the 12 years planning and implementation of the rural development initiative – the case of the 'Heritage trails through Dolenjska and Bela krajina in SE Slovenia'. Secondly, it is an example of sustainable rural development (which included cultural and heritage, vinicultural and gastronomic as well as ecological tourism) using an integrated approach in terms of startup, implementation and development, and benefits from a core of multiple stakeholders. Thirdly, the text seeks to show how practical experience is needed alongside theoretical concepts in order to identify the conditions necessary for success and to re-evaluate and improve development models. Finally, this case study reflects the category of 'nascent tourism in non-western societies', as described in the introductory chapter of this book by Prof. Svalastog.

One of the beneficial methodologies for growing and developing a level of tourism which is sustainable and enhances the totality of local and regional environments is a multi-stakeholder approach to tourism development. It is clear that:

- Entrepreneurship and entrepreneurial skills, harnessed in a bottom-up model of development, will have a huge impact on rural and agri-tourist micro-economies at a local community level. The effect in driving wealth creation and expanding employment can be measurable in a very tangible and transparent way.
- Furthermore, multi-stakeholder tourism projects help in the process of ownership transformation, by forcing *public, **private and ***social agents and enterprises to work together for the common benefit. Because of the bottom-up approach the measurable value at an enterprise or agency level is more tangible and obvious.
- We can also see that by engaging local public agencies, the system of environmental planning and protection can be assured. In this way the sustainable nature of tourism and its impact on the local environment can be evaluated and given due priority.
- At the same time, in such integrated projects, individual entrepreneurs begin to comprehend and understand the value of co-operation as well as competition. A key feature is often the need for small-scale tourism entrepreneurs to develop a promotional mechanism to market their product or service at a wider national and international level. Individually, the costs of such an activity

 Tourism as a Resource-based Industry: Based on the work of Sondre Svalastog (eds A.L. Svalastog, D.K. Müller and I. Jenkins)
DOI: 10.1079/9781800621480.0011

are too great for micro-enterprises, but they are possible for groups of enterprises. This demonstrates how an integrated model enables participants to benefit from the totality and complexity of resources and skills held by all stakeholders.

Clearly, the model, as demonstrated in this case study, has a very precise local/regional orientation. The emphasis is on a broad and contextually sensitive resource base, which helps to identify and explain how development and sustainability can be achieved. It also has a strong multi-stakeholder approach which in many ways illustrates the impact of EU-funded programmes on the concept of subsidiarity, which aims at seamless connectivity between EU supranational policy and funding, member state objectives in macro-economic harmonization and stabilities and local micro-economic needs. (The principle of subsidiarity is defined in Article 5 of the Treaty establishing the European Union and was intended to ensure that decisions are taken as closely as possible to the citizen and that constant checks are made as to whether action at supranational level is justified in the light of the possibilities available at national, regional or local level. The Edinburgh European Council of December 1992 issued a declaration on the principle of subsidiarity, which lays down the rules for its application (European Parliament, 2007).) We can therefore see that the local level becomes more and more important. Also, in the light of the recent and continuing pandemic of Covid-19 (as at 2021), people are seeking unique and unspoilt destinations where they can relax in quality and green environments, together with enjoyment of local traditions and everyday life in the local communities. To access and enjoy such attractive environments and services we need responsible, local management. Therefore, local authorities are generally responsible at the subsidiary level, which is below national and regional government provision of social services. At the local level it is possible to create micro-economic infrastructure, regulate local business activity and manage the natural environment. The local level has many developmental instruments at its disposal. Yet in addition to these roles, perhaps the most important is that of facilitator among the diverse public, private and voluntary interests seeking to influence the direction of local development. Only with such a facilitator, be it an interdisciplinary team or an individual, can a community of diverse interests define a shared vision and act consistently with that vision. At the local level, sustainable development is achieved by steering local development activities to simultaneously achieve three objectives:

- increased local social welfare;
- greater, and more equitably distributed, local economic wealth; and
- enhanced integrity of local ecosystems.

(ICLEI, 1999)

The Dolenjska–Bela krajina Heritage Trail Case Study

Research and methodological platforms

Tourism has been strongly interwoven into various policies which have emerged from various international agencies, such as UN programmes, as well as from the institutional structures of the European Union (Council, Commission and Parliament). The EU is seen to have impacts on employment, the environment, rural development, urban regeneration and sustainability. Importantly, the concept of bringing together various stakeholders has been an important and powerful element in the concept of sustainable rural tourism development.

There are important future elements that can be identified with this case study and these are presented below. The heritage trail model can be seen in the context of current EU-wide initiatives that seek to enhance the common cultural heritage, through the creation of a European heritage label. Cultural heritage, as we will see in the case study, also includes the protection of traditional local culture and, in particular, endangered folk arts, crafts, trades and knowledge. Tourism is one of the many external forces influencing the direction and options for local development. The question of whether tourism can be sustainable – that is, whether it can contribute to local sustainable development – is therefore rightfully addressed. An honest and practical discussion on sustainable tourism must take place with the communities that are being influenced by tourist industry development. It must create accountability of the tourism industry in unison with locally defined development visions. The true proof of 'sustainable tourism' will be the sustainable development of local communities that serve as tourist destinations (Koscak and O'Rourke, 2021). On the other hand, appropriate market segments must be targeted, when marketing the heritage destination. Prof. Svalastog (Svalastog, see Chapters 1–5 in this book) proposed expansion of Aubert's typology of tourism types who are likely to visit the destination. We believe that the 'outdoor' and the 'expert/connoisseur' are two types of tourist who would probably be most attracted to a heritage destination and need to be targeted when marketing the destination.

A further key element, related to the variety of stakeholders in sustainable tourism development, is the role of small to medium-sized enterprises (SMEs) and particularly micro and smaller-sized enterprises. EU institutions share a common view that SMEs/MSEs should have greater involvement in tourism-related entrepreneurship and that possibilities for simplifying procedures to obtain financial support should be found. Two additional primary requirements are postulated in the text: first, the need for a local focus and a bottom-up approach to development; and, secondly, the need to review failures resulting from a lack of professionalism, understanding and capacity. These can be addressed through the establishment of properly supported multi-stakeholder destination management structures, with links to relevant regional and national bodies. The text below proceeds to provide evidence and detail to support this approach through reference to theory and practice (Koščak, 2002).

Another important element is to study the impacts of tourism upon host communities as well as upon host cultures. Tylor (1871), showed that 'culture or civilisation, taken in its wide ethnographic sense, is that complex whole which includes knowledge beliefs, arts, morals, law, customs, and any other capabilities and habits acquired by man as a member of society'. Malinowski gave a more universal and less anthropocentric definition, namely, 'culture is a functioning, active, efficient, well-organised unity, which must be analysed into component institutions in relation to one another, and in relation to the environment, man-made as well as natural' (Malinowski, 1931). In the report on Wales in 1983, the word culture was taken to mean 'the system of values, beliefs, behaviours, morals, and other social phenomena shared by a group of people, based on their common experience of life, language and history' (Jones and Travis, 1983).

In order to assess these important components of sustainable tourism development, a detailed carrying capacity study was part of the project activities. Carrying capacity can be defined as 'the maximum number of people who can use a site without an unacceptable alteration in the physical environment and without an unacceptable decline in the quality of the experience gained by visitors' (Mathieson and Wall, 1982, p. 21)

The factors that need to be considered are:

- physical impact of tourists;
- ecological impact of tourists;
- tourists' perceptions of overcrowding; and
- cultural and social impact on local people.

The heritage trail is intended to be sustainable:

i. commercially successful (and therefore profitable);
ii. composed of local, small-scale enterprises – integrated into the local economy and contributing to 'locally defined social, economic, environmental and cultural needs and objectives';
iii. respectful of the environment and designed to minimize environmental impacts through the use of existing structures and appropriate technology;
iv. locally defined, owned, planned and controlled.

The carrying capacity study is central to meeting these objectives; it was necessary in order to identify environmentally and culturally sensitive areas and ensure that the heritage trail(s) is sustainable.

The purpose of the carrying capacity assessment is to ensure that tourists and day visitors attracted to the heritage trail sites will not have a deleterious impact on the cultural or natural sites, that overcrowding will not result in visitor dissatisfaction, and that local people will not feel antagonistic towards their 'guests'. This is essential if tourism is to contribute to the conservation of cultural and natural heritage though the realization of economic value and raising awareness of, and commitment to, the local patrimony.

Carrying capacity (CC) can be measured in a number of different ways, but in the development of heritage trails the major issues are concerned with the development of the local economy and the management of sites and whether or not particular sites should be excluded on the grounds that the likely impact of additional visitors will damage the site or be unwelcome to local people.

The first stage of the carrying capacity work is to begin the process of assessing the carrying capacity of each of those sites being considered for inclusion in the trail. After the visit of the CC specialists, a series of further social carrying capacity studies should be conducted and this will also begin the process of participatory planning.

Local people must be consulted in the assessment of landscapes and cultural and natural heritage assets. It is essential to ensure that the local impact of increased heritage tourism is brought within the process of developing and marketing tourism products.

The case study

The rural case study presented is one of a region in Slovenia along the border with Croatia (Fig. 11.1), where we tracked a 15-year process, from preliminary idea to the operational reality of sustainable international tourism in a strategically located destination-region (Koščak, 1999).

Origins and catalysts

The 30-year period from 1960 to 1990 saw distinct phases of evolution in tourism, planning, and conservation-focused thinking and action in the western world. This led to the concepts and processes of sustainable tourism planning. For example, in the UK, by the end of the 1980s, a National Task Force on 'Tourism and the Environment' had been established in order to provide sustainable tourism guidelines for three problem categories:

- the countryside
- heritage sites
- historic cities and towns

In the case of this Slovenian example, an additional factor is the multiple dynamic of international, national, regional and local agencies involved in the project. These were drawn from public, private and social sources, but the key actors and catalysts who can be identified in this story were the Slovenian Ministry of Agriculture, the Bavarian State Ministry for Agriculture, the Faculty of Architecture in Ljubljana, the European Commission's Tourism Directorate, a regional chamber of commerce, a commercial tourism operator and, at a later date, an international market research consultant. The case identifies and discusses a number of general and more specific themes, trends and other issues that

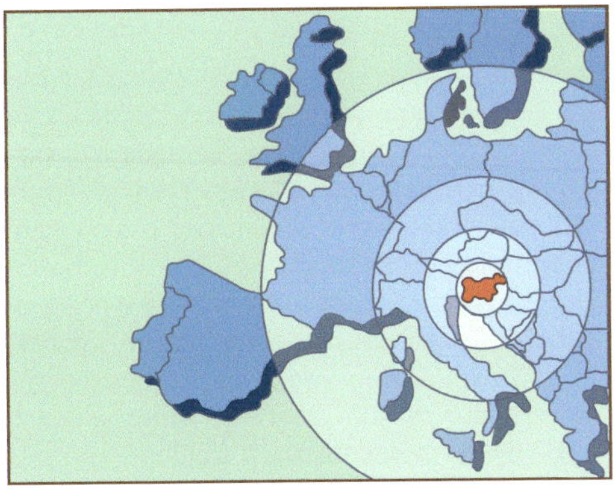

Fig. 11.1. Geographic position of Slovenia in the EU and the region of Dolenjska and Bela krajina in Slovenia. (Figure courtesy of the author)

present challenges and opportunities for rural tourism related to local culture and heritage. Some of these relate to market conditions and patterns of visitation and spending, including the trend towards all-inclusive models. Forms of the rural cultural tourism product are identified and factors affecting their development are discussed. Of most significance, however, are the issues arising from reduced public sector resources, the particular needs of small tourism enterprises within a challenging macro-economic environment, and the importance of inclusive and supportive structures for stakeholder participation that provides leadership and co-ordination within local destinations (Koščak and O'Rourke, 2021). This process has been identified by Prof. Svalastog's thesis in his introductory texts (see also Chapters 1–5).

Model of Integrated Rural Community Development Project in Slovenia

The relatively rapid stabilization of the economy, and Slovenia's appearance as an island of political and economic stability on the edge of the Balkan madhouse from 1991 to 1995, assisted Slovenia's tourism beginning to drive a highly marketable-oriented tourism development policy. As a result, significant studies were carried out by the Ministry of Economy. The overall strategy, which began to unfold with the support of various EU pre-accession and candidate country programmes, had two main strands – the local and regional, and the national. The model was evidently driven by:

- willingness to bring a number of stakeholders into the development and implementation process – domestic and international, private and public;
- the decision at a relatively early stage to launch the Integral Development of Rural Areas and the Renovation of Villages project, known by the acronym CRPOV. This project was backed by the Ministry of Agriculture, Forestry and Food, which played a very important role and at one time involved more than 250 micro-projects throughout the country, primarily aimed at possibilities of development for rural economic diversification.

The CRPOV programme (Integrated Rural Development and Village Renovation, 1996), which commenced in 1990, was also associated with the UN Food and Agriculture Organization (FAO) and the Bavarian Ministry for Agriculture. Bavaria helped in the initial phase, transferring experience and know-how. CRPOV was based on a bottom-up approach involving an initial 14 local project areas, starting in 1991. Two of the project villages were located in the Slovene municipality of Trebnje with around 500 local residents involved.

The community development role of CRPOV involved many local village meetings linked to the economic need for diversification of the rural economy. CRPOV worked together with an expert team on strategy and action. Critically, this case study relates to a rural region which sits strategically between Ljubljana and Zagreb, on the international motorway from Belgrade to Ljubljana. This has an important location potential for selling locally sourced food and wine products as well as craft and tourism products. Tourism in this area is based on the appeal of a gentle landscape of hills and river valleys for walking, horseback riding, cycling, angling, rafting, or the simple enjoyment of its unspoilt character.

The CRPOV, as an Integrated Rural Community Development programme, led the way towards rural product development, and as a by-product, community-based sustainable tourism. Such tourism requires partnership and co-operation between the public, private and NGO voluntary sectors. Co-operation of this sort was not common in the period 1992–1995 in Slovene tourism. It was clear, however, that sustainability, in Slovenia or anywhere else, requires community involvement together with the firm commitment of local actors and producers of products and services. The appeal of such development is to add tourism products to the other rural products, which they complement. Community-based rural development is thus an ideal starting point for sustainability, whether in agriculture or tourism. This creates an 'environment' in which new opportunities for economic diversification, job creation, value added to agricultural products, local guiding, and new farm services can occur. In this process, institutions like an agricultural extension service and others play a very important role in terms of capacity building and human resource development (Koščak, 2012).

International Team Heritage Trail Consultancy and Methodology

The background to this CRPOV programme as well as the parallel development of wine trails, prompted the Regional Chamber of Commerce of Dolenjska and Bela krajina to accept an invitation by a consortium (which had, in 1996, secured European Union funding to launch two pilot projects in Slovenia and Bulgaria) to create heritage trails. The consortium included Ecotourism Ltd (a British consultancy firm), PRISMA (a Greek consultancy firm) and ECOVAST (European Council for the Village and Small Town). All of these were supported by regional and national institutions in the categories of natural and cultural heritage.

Participative methodology

The UK/Slovene Heritage Trail team conducted a 'Tourist Resource Inventory & Selection', based on natural, built and living cultural heritage resources in the selected region. Some 150 sites were identified and proposed by the different partners involved in the participation process for the heritage trail. From this large number, 28 sites were selected, to be networked in a trail system for the area. The idea was to develop a tourist product which was capable of offering opportunities for stays of up to seven days in the region. Two key access forms were used for the clustering of attractions, one a 'flower structure' and the other a 'garland structure'. Existing tourist assets and potentials were the basis of these groupings.

Key results

A major result of this work was the creation of a regional partnership of 32 organizations, from the public, private and NGO sectors, which signed an agreement to co-operate in the heritage trail's implementation phases of marketing and product development. This partnership, working under the umbrella of the Regional Chamber of Commerce, was in operation for 12 years and was, in 2009, transformed into the LEADER Local Action Group (LAG), responsible for overall rural development in the region of south-east Slovenia, including sustainable tourism. The partnership supports, co-ordinates and brings together the provider partners. Work, in general, consisted of marketing activities, product development and training activities, where different combinations of partners, institutions and individuals are involved.

For marketing purposes, a local commercial partner, Kompas Novo Mesto, was invited into the partnership in 2001 in order to articulate a stronger and more effective reach on foreign markets. Kompas was engaged to act as the marketing agency on behalf of the heritage trail partnership. Kompas Novo Mesto was also responsible for the statistics of visits to the Heritage Trails, which they provided every year from 2002 (see Fig. 11.2). Although the official launch of the product was in 1997, at the World Travel Market in London, followed in 1998 by a presentation at ITB/Tourist Fair in Berlin, there was no significant response. Foreign markets at that time had limited awareness about any Slovene tourist products, other than what can be described as the constantly featured traditional Slovene tourist icons such as Lake Bled, Kranjska Gora ski resort, Postojna cave and Portoroz seaside resort. Slovenia was mixed with Slovakia, Slavonia (region in Croatia) etc., and it was only more fully recognized after it become part of the EU in 2004. This created a lot of challenges for marketing of Dolenjska and Bela krajina heritage trails, particularly in the first years of the commercial launch.

The effective commercial launch, in London, of the heritage trail at an international level was delayed until 2002. The launch included a foreign tourist industry adviser and a more professional approach. This World Travel Market launch had the active support of the Slovenian Tourism Board, together with other relevant institutions.

Stages of Commercial Product Adaptation and Implementation

Despite the launch of the heritage trail in the domestic market, followed by the international launch at the World Travel Market in 2002, the level of response by foreign tour operators and travel agents was weak. It became clear that external help was required to target appropriate foreign tourism trade partners as well as in identifying and selecting niche markets.

From the market research conducted on Slovenia's key foreign markets, the special interest markets, with a focus on either cultural tourism or nature tourism (ecotourism) were selected. Independent and some major commercial operators were approached by phone, fax or online. Two hundred firms were identified in seven European countries. Of these, 60 firms were contacted by at least two contact modes, but only six showed some degree of interest (Travis, 2002).

The low response revealed that there is much interest in Slovenia as a high-growth destination country, but the international industry identified Slovenia as having three major attractions – the 'tourism icons' already mentioned – lakes and mountains, caves and sea. For a significant period of time, Slovene overseas marketing had tended to focus only on these well-known destinations! This changed significantly in the last decade, when Slovenia became known for its green and sustainable development policy in the tourism sector. As Svalastog notes, one of the biggest challenges is how to integrate tourism to become part of an effective driver in local communities, and what he calls *resource-based tourism*. This is the segment of tourism that is governed by *recreational opportunities, activities and experiences* in a given area, particularly in rural parts of the country.

By 2003, low-cost airlines made Slovenia easily accessible to high-spend markets. Air travel cannot be a basis for sustainability but may have to be used as the initial opening-up phase for a new destination or product development. The current agenda focusing on climate change does not favour this mode of transport for the future. Tourism, and its transport sector, is one of the main polluters and will need to adapt future strategies, as discussed in COP26 in Glasgow (Glasgow Science Centre, 2021). Ultimately, access using unified rail travel must be a long-term primary aim. Train travel may be romantic if one chooses the right train; car travel also if it is combined with stops *en route* in some special eating spots. These modes of transport may create new possibilities and a unique experience for visitors, enabling them to enjoy their stay holistically and safely. This still needs to be achieved in the case of Slovenia, which is still lagging behind in its railway sector. However, the opening of the heritage trail market was via the low-cost airline destination airports of Ljubljana (Easyjet), Klagenfurt (Ryanair) and Graz (Ryanair), with ground transport routing via Ljubljana. Detailed discussions with key operators by phone showed that there were two viable special-interest packages, which could appeal commercially:

- a heritage trail add-on package for Bled (lakes and mountains) or Ljubljana (city and culture); and
- an integrated new 'highlights of Slovenia' holiday, which started with 25% of the time at two existing iconic destinations (Bled and Ljubljana) and the remaining 75% of the time on the heritage trail.

Testing of this product with a group of six UK travel professionals was extremely successful. A second tour with tour operators from Germany and the UK in 2004 was less successful. In 2005, a specialist walking-tour firm assembled its bespoke and individualized heritage trail offer and, as a consequence of the visit, independent tour operator firms were preparing for launching, online, two individualized alternative packages.

These individual packages were, and still are, marketed by regional DMOs (destination management organizations) and can be seen at: https://www.visitdolenjska.eu/en/heritage-trails.html (accessed 19 April 2022).

Heritage Trails and Creative Industries in Case of Dolenjska and Bela Krajina Development

At a very early stage of heritage trail development the project produced a number of initiatives which supported and encouraged individual and private sector companies to become an important part of this development. Some major ideas behind this were to create opportunities for new jobs and economic diversifica-

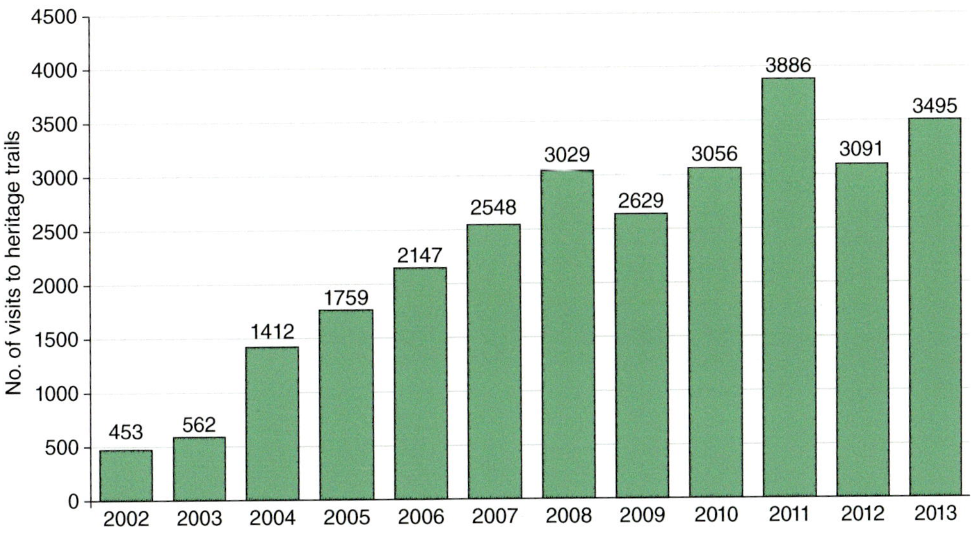

Fig. 11.2. Volume of visits to heritage trails product, 2002–2013. (From KOMPAS Novo mesto, 2014)

tion in rural parts of Dolenjska and Bela krajina, the south-east region of Slovenia, which is very similar to the Norway model, as described by Prof. Svalastog in his introductory chapters. DMOs are rethinking their delivery on how to serve the needs of local populations, repositioning their websites as information portals for residents. They update what restaurants are serving, what is being delivered and their takeaway foods. Furthermore, they are also offering advice on loans and business support, or connecting residents to charities and organizations. Another service deals with helping the most vulnerable in their communities who are in need of urgent support. A result of this rethinking means that destinations that are different and previously unknown could become more marketable. There are many of these areas which are not crowded with tourists and it is the DMOs who have this information rather than other types of tourist organizations and agents. If local DMOs are able to conduct some research and communicate their findings, they may also be able to impress their clients with unique knowledge and offers (Koščak and O'Rourke, 2021).

With such initiatives and support, the heritage trail partnerships are able to provide funding. Also, some 600 individuals took different types of education and training, such as meat and milk processing, bakery, beekeeping, wine production and its marketing, tourist guiding, traditional fruit drying and many others. All these individuals received certificates, allowing them to open individual businesses, satisfying all legislative requirements and also enabling them to apply for further funding from the rural development programmes, offered as a result of Slovenia joining the EU in 2004.

Different local thematic routes, such as wine, fruit, cheese and others, were created where local entrepreneurs started to create new tourism products and, through the marketing of heritage trail partnerships, offered them to domestic and international markets. All of the above-mentioned activities were conducted and implemented by heritage trail partnership institutions, such as the Chamber of Agriculture, which was responsible for the organization of training and certification. These certificates were based on the national curriculum for supplementary activities and Regional Development Agency (RDA). The RDA offers support and expertise in providing know-how on business plans and other entrepreneurial activities together with information on how to apply for tenders for various EU funding. This therefore opens the door to a number of new opportunities including the following:

- slow tourism;
- the experience economy;
- multi-sensorial products; and
- responsible local destination management.

(Koščak and O'Rourke, 2021)

This initial stage of certification was important in order to ensure that businesses operated legally and in accordance with new EU regulations and requirements. In the next stage of development more innovative and robust initiatives took place. Some individuals and even a group of partners decided to develop new products based on past traditions, giving them a fresh new outlook reflective of modern EU tourism markets.

Finally, on many of the new developments, tourists were already present, which 'brought' added value to the rural economy of Dolenjska and Bela krajina. One of the most popular and high-selling products, besides farm tourism B&B, was 'vineyard retreats', which is renting traditional wine cellars to guests as self-catering units (Fig. 11.3). Also successful in rural south-east Slovenia were attractively located buildings, which provided the potential for a unique and unforgettably rich experience, by providing rural accommodation, traditional food and opportunities for active holidays.

Learning points

1. It is evident from the case study that the 'heritage development for tourism' phase was preceded by the work on integrated rural community development. This stimulated a community-based approach to development in which context tourism was a part of the economic mix. This created a real expectation that sustainability might be achieved, via the local communities' support for a new mixed economy. The project indicates that sustainable development can underpin successful tourism planning if the correct strategy is chosen.

2. Evidence from the project has also made clear that heritage-resource-based tourism development, if it is to be sustainable, must (i) show respect for the carrying capacity of resource zones, be they robust or fragile; and (ii) have rural community involvement and commitment to tourism, providing a stake in it and deriving net gains from it.

Fig. 11.3. Vineyard retreats. (Photos courtesy of the author)

3. Successful tourism development arises because the destination creates potential tourism products from which the community benefits economically. Rural tourism products have to be adjusted to fit niche market demands as these products are highly competitive sectors internationally. Thus, market knowledge and understanding must be built in early to the development process or it becomes far more difficult to be successful.

4. New tourist destinations are very difficult to launch internationally, even if they have high accessibility, unless they can be linked and tied into existing tourism icons or magnets. This new Slovene offer had to be adjusted to do precisely that.

5. The 'gateway' identification is critical in new product formulation, whether this be a selected airport, seaport, railway station or whatever. If the gateway is the airport of an attractive heritage city (such as Ljubljana), then an add-on package is possible, as well as links to a popular 'short city break' destination, which creates added value.

6. Continuity of personnel in a development process is of real importance. The role of the project manager in initiation and continuity is critical, and the continuing interaction with external partners who are supportive and who share a belief in the integrity of the development over the long term is also valuable.

7. This model, ultimately, is one of community-based multiple stakeholders, having the equal support of small rural operatives and major agencies. The support from several levels: local, regional, national and international, have enabled the 13-year development cycle of the Dolenjska and Bela krajina heritage trail project to achieve its aims.

8. This project was designed and implemented from 1996 to 2009, supported by local, regional, national and international bodies. It involved the networking of carefully selected cultural sites and experiences in a little-visited but highly attractive part of Slovenia, in the form of a trail that could be followed by groups and individuals using various transport modes. The study explains the challenges faced in gaining international market interest and how these were overcome through establishing linkages to existing markets and gateways. A legacy of the project has been the creation of new products, e.g. in food- and drink-based experiences and heritage themes, together with renewed destination identity and networking. The result has led to rural development, regeneration and diversification to complement existing tourism products.

The above points of interest were identified by Prof. Svalastog in his model and thesis, supporting the concepts that he discussed.

Critical success factors

There are good reasons why the Slovene heritage trail model has been successfully adopted in several neighbouring countries as an initiative for rural regeneration through sustainable tourism, namely:

Factor 1 – Economic regeneration

A heritage trail is created as a tool for rural economic regeneration. The heritage trail extends tourism from existing centres into new and under-visited areas, by increasing the number of visitors, extending their stay, and diversifying the attractions and services offered to them: expansion, extension and diversification.

Factor 2 – Contributing to regional tourism development

The heritage trail is a tourism product which makes the natural and cultural heritage of a region the focal point of the tourist product. The development of such a product is, therefore, an integral component of the development of the whole region as a tourism destination. However, a heritage trail is only one product, and many regions have other tourism products on offer which may not be included in the trail. In creating heritage trails in Slovenia, there was frequently a temptation to include all tourism attractions and services in the region. But to give in to such a temptation would have been to lose the focus of a well-defined tourism product.

Factor 3 – Complementing other tourism products

Although a heritage trail focuses on only some of the attractions of a region, it can be complementary to other tourism products on offer. For example, it can contribute to economies of scale in regional promotion – in Slovenia, the heritage trail and spa tourism were promoted jointly, and costs of this shared. A heritage trail can also contribute to a wider choice of products for target markets. Taking the example of Slovenia again, spa tourists may be interested in the heritage trail product, and heritage trail tourists may enjoy the spa facilities.

Factor 4 – Transferability

The heritage trails concept is transferable to other regions and countries where there is sufficient natural and cultural heritage to attract tourists and where there is a local desire both to benefit from tourism and to safeguard that heritage. This is particularly the case in parts of central and eastern Europe where established settlement patterns and rural economies have developed in similar fashion to those in Slovenia.

Factor 5 – Sustainable tourism

A heritage trail focuses on the natural and cultural assets of a rural region. This runs the risk of exposing some of the most vulnerable sites in a region to excessive numbers of tourists. The preparation of a heritage trail, therefore, must include a tourism 'carrying capacity study' at each proposed tourism site. If a sudden increase in tourists risks damaging the physical or natural attributes of a site, or if it were to exceed the tolerance of the local people, it should not be included in the heritage trail, until preventive measures can be implemented.

Conclusion

Tourism has been recognized as a global industry – one of the largest industry sectors in the world. As with any global industry, tourist business activities may have considerable impact on local development trends. The local impacts of the tourism industry are diverse and are often unique to the tourism sector. Tourist activities, as traditionally defined by the tourism industry, fundamentally involve the transportation and hosting of the tourism consumer in a local community, i.e. 'tourist destination', where the tourist product is consumed. No other global industry structures itself in such a way that the consumer is brought to the product, rather than the product being delivered to the consumer in his or her own community (ICLEI, 1999).

Tourism activities can, in particular, degrade the social and natural wealth of a community. The intrusion of large numbers of uninformed foreigners into local social systems can undermine pre-existing social relationships and values. Tourism in natural areas, euphemistically called 'ecotourism', can be a major source of degradation of local ecological, economic and social systems. The intrusion of large numbers of foreign visitors with high consumption and possible high waste into natural areas, or into towns with inadequate waste management infrastructure, can produce changes to those natural areas at a rate that is far greater than that caused by local residents.

Solutions to adverse tourism impacts are to be found in the shared interest of local communities, tourism businesses and tourism consumers, and their ability to maintain the natural wealth and social heritage of the tourist destination. In the first instance, therefore, an institutional mechanism must be established, relative to each destination, to articulate and develop this sense of shared interest. To secure an acceptance of these mechanisms, the participation of all interested local groups or interests must be guaranteed. Dialogue must take place in an open and transparent way. Experience demonstrates that if dialogue among interested parties is to have a real impact on development it must generate accountability among these parties with regard to future investments, practices and policies. Consensus or decisions arising from dialogue must be reflected in institutional action. Only through such accountability can individuals, businesses and organizations maintain a commitment to continued dialogue and a common agenda for local tourism development (ICLEI, 1999).

Notes

* by public we mean municipal/local government, state agencies and international organizations operating in a local or regional framework.

** by private we mean privately owned companies, including quoted or unquoted companies, as well as partnerships or self-employed individuals.

*** by social we mean entities established for mutual benefit, including co-operatives, societies and not-for-profit agencies.

Literature and Bibliography

European Parliament (2007) Resolution of 29 November 2007 (Commission recommendation 2006/2129-INI) on a renewed EU Tourism Policy: Towards a Stronger Partnership for European Tourism. Available at: https://www.europarl.europa.eu/doceo/document/A-6-2007-0399_EN.html (accessed 8 July 2022).

Glasgow Science Centre (2021) What is COP26? Available at: https://www.glasgowsciencecentre.org/discover/our-world-our-impact/what-is-cop26?gclid=Cj0KCQiAzMGNBhCyARIsANpUkzMzyVJLW-wl_3RCzyuB2o8szEDZcvqAPLr9F7I legkm6TF4QjyNMLaRMaAiyEEALw_wcB (accessed 16 November 2021).

ICLEI (1999) Local governments for sustainability. Available at: https://iclei.org/en/Home.html?gclid=EAIaI-QobChMI4ejx_r- q7glVjuR3Ch3FFwKAEAAYASAAEgJ2HfD_BwE (accessed 16 November 2021).

Integrated Rural Development and Village Renovation (1996) Development programme, Ministry of Agriculture of Republic Slovenia, Ljubljana.

Jones, A. and Travis, A. (1983) *Cultural Tourism: Toward a European Charter*. Report to WTB, EW Tourism Consultancy, Birmingham, UK.

KOMPAS Novo mesto (2014) Unpublished statistical data. Novo mesto.

Koščak, M. (1999) Transformation of Slovene rural areas along the Slovene-Croatian border. PhD thesis, University of Ljubljana.

Koščak, M. (2002) Heritage trails: Rural regeneration through sustainable tourism in Dolenjska and Bela Krajina. *Rast* XIII 2(80), 204–211.

Koščak, M. (2012) Po poteh dediščine – od teorije k praksi, Priročnik za načrtovanje trajnostnega razvoja in turizma z vključevanjem naravne in kulturne dediščine s praktičnimi primeri. Studio MKA Ltd., Self-published Guidelines, Trebnje.

Koščak, M. and O'Rourke, T. (2021) *Post-Pandemic Sustainable Tourism Management: The New Reality of Managing Ethical and Responsible Tourism*. Routledge, Abingdon, UK.

Malinowski, B. (1931) Culture. In: *Encyclopedia of the Social Sciences*. Macmillan, New York, Vol. 4, pp. 621–645.

Mathieson, A. and Wall. G. (1982) *Tourism: Economic, Physical and Social Impacts*. Longman, Harlow, UK.

Travis, S.T. (2002) Market Research on Slovenia's Key Foreign Markets. Dolenjska and Bela Krajina Marketing Association, Chamber of Commerce, Dolenjska and Bela Krajina.

Tylor, E.B. (1871) *Primitive Culture: Researches into the Development Mythology*. HarperCollins, New York [reprinted].

UNEP/ICLEI (2003) Tourism and Local Agenda 21 – Role of Local Authorities in Sustainable Tourism. United Nations Environmental Program, Paris. International Council for Local Environmental Initiatives, Freiburg, Germany.

12 The Post-industrial Urban Landscape: Heritage Resources Enhancement in Urban Renewal – the Case Study of Central Birmingham, UK

Anthony S. Travis

Introduction and Aim

The case study of central Birmingham's transformation over a 30-year period, from the 1970s to the first decade of the 2000s, is an extraordinary one, relating to the death of a great industrial city and its reinvention, or recycling into a post-industrial city of great quality.

It is the intention to tell this story of complex economic development changes with the use of a range of responsive instruments, to transform the physical infrastructure systems and the associated building superstructure, with varied strategies employed. The results of this complex process, or production, has been to create an urban environment of great quality, in which inherited resources, both natural and man-made, have been enhanced, and activity levels in the transformed spaces vastly increased. The focus is on how this happened, in which phases and with which detailed results.

Early 20th-century Birmingham was the 'Workshop of the World' – a city with traditions of non-conformity, trade, manufacture and innovation. It had been a 'city of a thousand little meisters'. By the 1960s, however, Birmingham was a tired and drab city, slowly recovering from World War II, reconstructing its war-damaged and blighted environments and needing massive industrial reinvestment. The launch of the Mini car's production there, in 1960, and the growth in car ownership, coincided with the Inner Ring Road construction and the start of the motorway building phase. Birmingham's central location, nationally, and its entrepreneurial traditions augured well for the future, but many of its small, old metal and jewellery industries were in decline. The opening of the Bull Ring Shopping Centre in 1964, with its siting on the Inner Ring Road, hailed the 'Motorway City', as devised by City Engineer Manzoni. However, 1960s central Birmingham was a down-to-earth '9 to 5' city, with dirty, inaccessible canals, a colourless centre, lack of trees and noisy, with a need for residents to go out to the countryside to find decent eating places and escape from the limited range of cultural activities in the city centre. It was a city losing its identity and in danger of losing its civic pride.

Though the 1970s started encouragingly with the opening of the new Rep Theatre in 1971 and of a grand new Central Library in 1973, the Middle East oil crisis soon led to industrial collapse in manufacturing. In the 1970s, 200,000 industrial jobs were lost and it seemed as if a great industrial city was about to die. It is, though, characteristic of this confident city that collapse led to positive and creative responses. By 1976, the National Exhibition Centre (NEC) was opened at the Solihull rail-entry point to Birmingham, and, also in the late 1970s, Project Aquarius brought the Department of the

 Tourism as a Resource-based Industry: Based on the work of Sondre Svalastog (eds A.L. Svalastog, D.K. Müller and I. Jenkins)
DOI: 10.1079/9781800621480.0012

Environment, British Waterways and the Birmingham Inner City Partnership together to start cleaning the canals, repaving the towpaths, opening up access to the canals, creating new bridges and interpreting and presenting the canals as a fine new resource for the city. These economic and environmental changes were the start of an economic and environmental quality renewal process that, over time, was to transform completely the city centre.

The key is that reinvestment had to go with a reorientation of the economy, life and values of the city, whereby existing natural resources had to be regained and enhanced and a city had to find vital new functions for its survival and well-being. It is significant that in the 1970s, too, the first conservation battle was won in Birmingham, when the fine Victorian Head Post Office building was saved and the principle of conservation of selected built heritage, together with new buildings, was established.

The Use of Responsive Instruments in the Production Process

The 30-year production, or transformation, process carried out in central Birmingham has involved the use of some six distinct instruments:

1. Economic developments have laid the foundation for changes, with a shift from a primary focus on manufacturing to a multi-pronged economy, with a growth of the service sector, particularly relating to exhibition, conference and congress sectors, retailing, catering and hotels, banking, insurance, accounting, education, research and development. By 1984, the city had opened its new international airport and, in the same year, a design evaluation was done for an International Convention Centre (ICC) to be built in the city centre. By the late 1980s, an Economic Development Strategy was using booster grants from the European Commission (£143 million) to complete the NEC and develop the NEC Arena (1980), fund part of the development of the International Convention Centre (ICC), develop the Aston Science Park and help fund central-area pedestrianization. The 1990s were to see key work done by the Birmingham Heartlands Development Corporation, training and business development by the Training and Enterprise Council (TEC) (£39 million), promotion of the finance sector by City 2000 and capital assembly for the development of four major building complexes: the ICC, Brindley Place Development, the Mailbox complex and, by the early 2000s, the second Bull Ring Centre, which opened in 2003.

2. Infrastructural systems planning saw the work on the canals – started in the 1970s – lead on to extensive waterfront developments by and during the 1990s, with the development of the pedestrian system across the city centre in 1992, which used the former trafficked streets and linked to a growing number of all-weather, undercover, pedestrian shopping malls. By 2004, Birmingham had already achieved the UK's largest continuous pedestrianized system across the whole of its central area. Furthermore, this was a landscaped and varied system with indoor and outdoor elements, waterfronts, attractions, shopping, rest and recreational elements, making it a mature system in its own right.

3. Building or superstructure realization had highlights in the opening of the NEC in 1976; the NEC Arena, added in 1980; the new international airport in 1984; and then the triple achievement in 1991 of the ICC, Symphony Hall and the National Indoor Arena (NIA). The ICC, in turn, generated the speeded-up development of the new Brindley Place district, plus, during the 1990s, the opening of the Mailbox complex and the construction of new hotels and restaurants. In 2003, the huge new Bull Ring Centre, which alone cost £500 million to create, was completed. Finally, in 2007, the historic 1830s concert hall, named the Town Hall, opened after total renovation. The expansion of the city's three universities over this period, to cater for more than 30,000 students, reinforced all these other major changes.

4. Perhaps most interesting of all was the evolution of **arts strategies** and **5. event strategies** in Birmingham in the 1980s and 1990s, as these were vital in changing the image of the city, or its 'rebranding' or 'repositioning' as marketers would call it. In 1980, Simon Rattle (now Sir Simon Rattle) was appointed conductor of the City of Birmingham Symphony Orchestra (CBSO) and set about bringing it up to world-class standard. Part of his strategy was the battle to obtain a new Symphony Hall with over 2000 seats and superb acoustics, a vision finally realized in 1991.

In 1986, Birmingham mounted a major bid to bring the 1992 Olympics to the city. It lost out to Barcelona, but so raised the profile of the city that huge inward investment followed. A feasibility study was done to see if a World Expo could be brought to the city. This all boosted economic development strategies and led on to arts strategies. The city's bid was to become a major cultural centre, which led to housing the CBSO in the new Symphony Hall, and bringing the Saddler's Wells Ballet from London in order to create the Birmingham Royal Ballet (BRB) in 1990. During the 1990s, the BRB's new home in the Hippodrome Theatre was to be totally rebuilt and it reopened in 2001. The opera component of the strategy was less successful.

By 1992, Birmingham had three venues for exhibitions, conferences, trade shows and indoor sports events – namely the NEC, the ICC and the NIA. It developed a unified management to market and manage all three great venues. The marketing was focused on Birmingham as 'Europe's Meeting Place' and the Birmingham Marketing Partnership, working jointly with the Birmingham Chamber of Commerce and Industry (BCCI), became a major player in the marketing and promotion field.

An Events Strategy – Towards the Millennium – in the 1990s changed the image of the city, catching up with the changing reality on the ground. In 1992, Birmingham was the UK's 'city of music'; in 1993 it had an 'Early Music Festival'; also by 1993, Birmingham City Pride was underway and efforts were made to get integrated urban strategies. In 1995, the city was voted 'UK's Cleanest City'; in 1996 it won the 'Britain in Bloom' competition; and in 1998, three major events took place in the city: the Eurovision Song Contest, the 24,000-delegate World Lions Congress and the World's G8 meeting – which caused the world's media to focus positively on Birmingham as an event venue for the first time.

There was great expansion of the hotel sector and an explosion of the catering sector (Birmingham now has over 200 restaurants, some Michelin-starred). By 2002, Crufts International Dog Show and the British Motor Show were at the NEC and Davis Cup tennis was at the NIA, whilst four major new tourist attractions were created in the city centre. These attractions were: the Sea Life Centre, the Ikon Art Gallery in Brindley Place, the 1992 Jewellery Quarter Museum, and the National Trust 'Back to Backs' Museum in 2004. All this, plus first-class shopping, meant Birmingham had become a major city-break tourist destination, in world terms, with impressive responses to its tourist marketing.

6. The **landscape treatment** of the resulting pedestrian realm of central Birmingham reflects and reinforces its transformation into a post-industrial city, where quality of surroundings – a safe, attractive and varied pedestrian environment, with choices of places to walk, sit, eat, drink, relax and be entertained and educated – are paramount.

Many resources have had to be transformed to achieve this. Former industry-serving canals have had to be cleaned and stocked with fish, their towpaths repaved, and new access points and interpretation boards created. Trafficked streets have had to have vehicular traffic removed and mature trees, seating, lighting, paving, sculpture, fountains and signs introduced as part of new 'pedestrian zones'. The ideas of a vibrant modern 'townscape', as first set out in the 1960s by Gordon Cullen to the architecture and planning professions, became the model for Birmingham.

The cleaning of the urban air is another resource change. Since the Clean Air Act became law, only smokeless fuels have been permitted in Birmingham and it now has very good air quality. Traffic removal has diminished urban noise, dirt and vibration, making the city centre not only visually improved but also quieter. The pedestrian system is over three kilometres in extent giving a choice of busy places and quiet places, with mood and activity-level variations over the course of a day, week and year. Saturdays are very busy, and farmers' markets, German Christmas markets and street festivals bring atmosphere and character.

Creating civic squares, such as Centenary Square, Chamberlain Square and Victoria Square, with fountains, sculptures and large spaces, allows for the city to have major street exhibitions and events such as New Year's Eve celebrations, concerts and dances in the heart of the city. The canal-sides have become lively places for café and nightlife, promenading and clubbing, while longboats on the canals offer trips and music events. Water taxis are also on offer. Thus, a city without a major river now uses its many miles of canals as a great recreational and amenity asset. Birmingham, as 'Europe's Meeting Place', is not only a great exhibition, conference and convention centre but is also a place where you now see the famous and the great of politics, music and theatre in its restaurants, shops, promenades and squares. The vision of a post-industrial city with a vibrant urban landscape has been realized.

Bibliography

Berg. J. (1997) *Positively Birmingham*. Birmingham Picture Library, Birmingham, UK.

BCC (1995) *Architecture and Austerity: Birmingham 1940–1950*. Birmingham City Council Planning and Architecture, Birmingham, UK.

BCC (1996) *Finding the Fifties: A 1950s Architectural Trail around Birmingham City Centre*. Birmingham City Council Planning and Architecture, Birmingham, UK.

BCC (1997) *Signalling the Sixties: 1960s Architecture in Birmingham*. Birmingham City Council Planning and Architecture, Birmingham, UK.

BCC (1998) *Shaping the Seventies: 1970s Architecture in Birmingham*. Birmingham City Council Planning and Architecture, Birmingham, UK.

BCC/Birmingham Convention & Visitor Bureau (BCVB) (2002) *Short Breaks in Birmingham*. Birmingham City Council, Birmingham, UK.

BCVB (2001) *One Stop – Birmingham, Venues and Services Directory*. Birmingham City Council, Birmingham, UK.

Birmingham Civic Society (2004) *Birmingham Heritage Buildings*. Birmingham Civic Society, Birmingham, UK.

Birmingham Heartlands Development Corporation (1994) *Annual Report & Accounts, 1993–94*.

Birmingham Marketing Partnership (2004) *Birmingham Visitor Guide* 2004/05.

Birmingham Training and Enterprise Council (TEC) (1994) *For a Brighter Future*.

Briggs, A. (1952) *History of Birmingham*, vol. 2. Oxford University Press, Oxford, UK.

British Waterways (n.d.) *Canals in Birmingham*. Cotterill, Cook, Birmingham.

Cherry, G.E. (1994) *Birmingham: A Study in Geography, History and Planning*. Wiley, Chichester, UK.

Chinn, C. (1994) *Birmingham: The Great Working City*. Birmingham City Council, Birmingham, UK.

Foster, A. (2005) *Birmingham: Pevsner Architectural Guide*. Yale University Press, New Haven, Connecticut.

Hatton Associates (n.d.) *Brindley Place, Birmingham*.

Hatton Associates (2001) *Think of a Place: Brindley Place, Birmingham*.

Heart of England Tourist Board (1998) *Case Study: Birmingham Waterfront*.

International Convention Centre Birmingham (ICC) (1991) *The Past, Present and Future – A Commemoration Book for the Opening of the ICC, Birmingham, 1991*.

NEC Group (1996) *Celebrating 20 Years of the NEC*.

13 Rethinking Tourism Development: Theoretical Perspectives – Concluding Synopsis

Ian Jenkins

Tourism remains a very important industry for many developed and developing economies. Svalastog's theory is as relevant today as when it was first written and the case study examples within the text illustrate and emphasize the benefits of the international tourism market. Indeed, the GDP for tourism in developed countries is, on average, between 7% and 10%. Norway certainly benefited from international tourism, even though since the 1970s it has declined. However, tourism maintains its position as a powerful economic driver and will remain so as the world's ever-growing population will continue to desire tourist experiences. The exponential growth of mass tourism during the 1960s and 1970s has continued into the 21st century, with the number of tourists reaching 1.4 billion per year (UNWTO, 2020a).

Theory Development

Since the publication of the thesis, the tourism market has significantly changed, yet much of the theory still remains relevant and applicable, producing an important perspective for today's tourism market and practitioners. For example, Svalastog's identification of tourism competencies connected to tourist typologies is fundamental to the development of tourism in today's competitive environment. The following remarks will encapsulate key elements in the thesis's applicability.

Tourist Types

Classification of tourist types has been embedded in tourism literature for some time and will inevitably evolve and change. However, within general tourism literature run common themes, classifying types of tourists and the different experiences they seek. The mass tourism markets of the 1960s and 1970s have splintered into many different variations even though mass tourism is still a worldwide phenomenon. There has been the growth of the 'adventurer', who at one time was regarded by many as a pioneer for the development of mass tourism. The adventure market has greatly expanded and is now a significant segment, including niche markets, with products including half-day adventures as part of mass-tourism packages. Certainly, a number of destinations can be recognized as significant locations for this adventure segment, perhaps two of the best examples being New Zealand and Iceland. Both Iceland and Norway are similar in having a high-cost economy, yet, since 2010, Iceland has experienced exponential growth, illustrating the notion of competencies that Svalastog explores within the book's earlier

DOI: 10.1079/9781800621480.0013

chapters. He introduces two new categories – activity tourists and utility tourists – which are certainly present into today's splintered market. In particular, activity tourism, has now transformed itself to become an adventure tourism segment which includes sports tourism. Probably the utility tourist is less obvious now, but could be connected to the new sharing economy, which has grown significantly.

Post-tourists and Overtourism

Not considered within this book has been the recent phenomenon of the post-tourist and overtourism. Overtourism narratives relate to 'honey pot' destinations such as Venice and Barcelona, where local residents have protested about the impacts tourism is having on their society and culture. Indeed, many of these overtourism destinations have businesses which are now run by corporate chains with few direct economic benefits for the locale. Many residents have moved out of immediate tourist destinations due to the often huge influx of tourists and corporate chains.

One consequence of the over-commercialization and acculturation of many popular destinations has been the growth of post-tourism and the post-tourist. This category complements Svalastog's classifications, being seen as a new tourist or anti-tourist who seeks destinations devoid of commodification. But is this not simply a reinvention of the explorer/adventurer classification? Certainly, the AirBnB market can be linked to this 'demassification' of the tourism product and an increase in what has become known as the 'sharing economy', whereby locals benefit directly from tourism or the authentic visitor. Post-tourism is overshadowed by impacts on the environment, another of Svalastog's key themes. Environmental issues were prominent in the 1980s but had not reached the crisis point in which today's tourism market finds itself.

Climate Change

The ever-expanding tourism market has increased its global carbon footprint to such an extent that it may be contributing up to 5% of global emissions (UNWTO, 2019). The exponential growth of aircraft is a large contributor to this and certainly the advent of LCCs (low-cost carriers) has increased demand and made many global destinations more accessible. A number of LCCs are rated in the top ten of global airlines (fleet size): EasyJet, Ryanair and South West being examples (Zhang, 2019). The IPCC (International Panel on Climate Change) issues regular warnings that current emissions, even after COP21 Paris, will seriously damage the global climate. This will inevitably impact on natural landscapes, changing the ecology of many areas and inevitably affecting numerous tourism destinations. This will require the development of new competencies for dealing with these effects, which must incorporate the skills of resilience and adaptation. Svalastog emphasizes the importance of natural landscapes to rural tourism, meaning that climate change will add a new dimension to local competencies as well as a need to diversify use of resources.

Rural Development

Norway is a country with magnificent landscapes and scenery whose physiognomy does not lend itself to large urban conurbations; therefore, central to much of Norwegian tourism is rural development. This is extensively discussed in the book and relates to how natural resources can be measured to identify natural capital, attracting tourists and, in particular, international tourists. Norway's high-cost economy was seen as an obstacle in attracting international tourists, especially given the cost of living and the high wages needed to pay employees in the industry. However, there are examples of countries that have overcome this particular tourism constraint. Iceland is an exemplar of how this can be done, as most of Iceland's tourism is based on nature (natural capital) and adventure tourism (Ferðamálastofa, 2016) However, Iceland's success is also linked to technological changes and the development of LCCs (as mentioned above) as well as its strategic position as a staging post between Europe and America. Whether local competencies were present is debatable, and although tourism did marginally contribute to the economy, this changed considerably from

2010. Iceland provides a useful example and realistic comparison with Norway and demonstrates how international tourism can be developed successfully in a high-cost economy. This supports, to some extent, the assertion of Svalastog in his thesis. However, tourism is a volatile industry and there are many examples of tourism leaving economies like Iceland 'high and dry' after years of success. Economic shocks can affect all sectors but tourism is particularly vulnerable. The 2020 pandemic of Covid-19 is certainly an example of this and has left the tourism industry badly affected, especially internationally (UNWTO, 2020b). This perhaps is a notion that could be further developed from Svalastog's theory, in particular how to incorporate or plan for global shocks.

Covid-19: Global Pandemic

Tourism is sensitive to economic shocks and there are many examples of how countries have been badly affected by this. The financial crash of 2008 greatly depressed global tourism in the same way that the 2020 pandemic has diminished global tourism markets (UNWTO, 2020b), with huge declines in many developed countries. In these countries governments restricted the travel of their citizens and their access to international tourism destinations and products. A global shock seems somewhat discordant from the main themes of this book, but it is suggested that many of Svalastog's themes are important to help deal with this crisis. Knowledge and competency are fundamental for adaptability and resilience to maintaining a viable tourism product. Many operators have been seriously worried about their viability and how to provide a product that tourists would purchase in the middle of a pandemic. Evidence from the UK is that although this crisis has seriously shaken the tourism market, consumers have still undertaken their holidays. There appears to be a change from an international to a domestic tourism market, reflecting patterns of the 1960s, and many rural tourism operators have been able to extend the season, in particular campsites and other similar establishments have extended their opening periods into late autumn (*The Week*, 2020). There is also evidence that consumers are using more local natural resources or just nature's capital in general. The important premise here is the ability of the tourism industry to adapt and be resilient to major changes in tourist patterns. It is suggested that disease outbreaks are not as uncommon as we think, and given globalization, 'disease competencies' need to be included in how to deal with economic shocks that affect tourism.

Competence

A central theme in Svalastog's work is that of tourism competence and engendering this within spatial locations and the economic system of a destination. This is very much an intangible aspect of tourism but with physical attributes. A fundamental element is that of skills that are not only related to the physical, such as carpentry, building and culinary, but also more intellectual skills such as entrepreneurship and creative imagination. Svalastog identifies that the more tangible aspects of the tourism product can be more easily skilled within the local population or region through training linked to education. However, the more theoretical aspects of tourism business and its delivery can be taught in higher education and associated courses. Hospitality courses have been the mainstay of many intermediate-level educational establishments and, globally, there has been an exponential growth in tourism and tourism-related courses. These have further splintered into more niche parts of the tourism industry such as adventure tourism, events and festivals, which are some of the more specialized courses currently being offered. It is to be questioned whether these university courses have delivered the competencies, and, in particular, the knowledge, which Svalastog identifies as essential to make a region or a nation more competitive in the international market for tourism.

Technology, Digitalization, Virtuality and Web3

During the time at which the thesis was written, there was a growth in the use of technology, which continues to play an extensive part in tourism development and products. Although Svalastog only mentions marketing incidentally as being the standard approach to encourage tourism to a destination, marketing has continued to play an important role in selling international tourism products. The internet, social media applications and other digital communication forms are now the foremost methods for targeting consumers. With the development of Web3 and further digitalization, the competencies for destination development have changed. The development has resulted in more visual and virtual forms of tourism, whereas Svalastog's discourse has a more fundamental and general focus on technology. Yet Svalastog acknowledged that competencies in technology were an important factor in achieving competitiveness. Also, technology is established within the knowledge concept of Svalastog's work and is therefore part of the knowledge delivery that universities and educational establishments will or should incorporate into their courses. There are now digital platforms allowing virtual journeys within a destination which must, to some extent, create a virtual experience that could not be experienced by tourists until very recently. As Svalastog notes, while reflecting on Porter's (1990) work, innovation does not just happen but is a nexus of a series of development processes and conversant knowledge. It has now become more or less essential for tourists to have some access to the internet or mobile/satellite networks whilst on holiday, and this technology has made even the remotest regions more closely linked to globalization and international tourism.

Product Development and Niches

A theme running throughout the earlier chapters in the book relates to niche tourism developments and their importance to the tourism market and its competitiveness. Tourists, in general, are much more informed and savvy consumers today, with increased access to knowledge and tourism destinations. The developed world also has a higher standard of living and expects quality of experiences, which must be matched by the quality of the tourism products consumed. A topical discourse of the early 2000s identified post-modernism as a factor explaining today's society, reflected in the splintering of products and markets, including tourism. A number of tourism authorities incorporated niche market development into their strategic documents; for example, Ireland and the British Tourism Authority. This seemed to reflect a change in the tourist gaze (as exemplified by Urry, 2002) at that time, yet very quickly the emphasis on niche markets seemed to evaporate and become moribund in terms of tourism objectives and marketing. The concept has remained, but rather more as a secondary consideration than of primary interest to tourist boards.

Svalastog links niches to the idea of general tourism products (see Chapter 5). Yet they appear as unspecified products, which is unsurprising as the literature on measuring specifically what niche tourism products are (Jenkins, 2007) is extremely sparse. Certainly, a unique cultural resource will provide a country with a competitive advantage given the right conditions and circumstances. Iceland is an exemplar of this and its tourism growth since 2010 has been exponential and based upon its unique natural landscape (including adventure tourism) and cultural heritage (Viking heritage). It has been a recent tourism success story and, in many ways, mirrors the discussions and assertions made in the formative chapters of Svalastog's work. There is also a certain irony, too, as this is, in a general sense, a Scandinavian country with a high-cost economy, very similar to Norway.

Concluding Remarks

It is clear that Svalastog's research and publications have contributed significantly to the body of knowledge of how tourism should develop in Norway. It can, no doubt, act as a muse for similar countries with small populations, mountainous regions, alpine conditions and high-cost economies. The case studies have illustrated how Svalastog's discourse can be materialized and developed to form a tourism product that can be internationally competitive, and he identifies some important principles which may help develop successful tourism at a destination. People create tourism, and the ability of the destination to develop competencies is an essential asset to ensure a destination's success in the international tourism market. Every country offers a number of unique characteristics but understanding the competencies of how these can be managed as a resource for encouraging international and domestic tourism is the challenge which Svalastog has left us. This challenge is constantly changing and materialized through a process of multi-dimensional resource inputs, woven into a space–time dimension at every global tourist destination.

References

Ferðamálastofa (2016) (Icelandic Tourist Board) International visitors in Iceland summer 2016. Available at: www.ferdamalastofa.is/static/files/ferdamalastofa/Frettamyndir/2017/januar/sunarkonnun/sumar-2016-islensk.pdf (accessed 6 November 2020).

Jenkins, I.S. (2007) Postmodern tourist niches: the case of literary festivals. PhD thesis, University of Wales, Swansea, UK.

Porter, M.E. (1990) *The Competitive Advantage of Nations*. MacMillan, London.

The Week (2020) Can an extended holiday season save the UK tourism industry? Travel and hospitality bosses hope to capitalise on post-lockdown sales boom. 7/9/2020. Available at: www.theweek.co.uk/tourism/108012/can-an-extended-holiday-season-save-the-uk-tourism-industry (accessed 6 November 2020).

UNWTO (2019) Tourism's carbon emissions measured in landmark report. Launched at Cop25, all regions. UNWTO, Madrid. Available at: www.unwto.org/news/tourisms-carbon-emissions-measured-in-landmark-report-launched-at-cop25 (accessed 6 November 2020).

UNWTO (2020a) International tourism growth continues to outpace the global economy. All regions. UNWTO, Madrid. Available at: www.unwto.org/international-tourism-growth-continues-to-outpace-the-economy#:~:text=1.5%20billion%20international%20tourist%20arrivals,in%20view%20of%20current%20uncertainties (accessed 6 November 2020).

UNWTO (2020b) World tourism barometer, May 2020: Special focus on the impact of COVID-19. UNWTO, Madrid. Available at: www.e-unwto.org/doi/epdf/10.18111/9789284421930 (accessed 6 November 2020).

Urry, J. (2002) *The Tourist Gaze*. 2nd edn. Sage, London.

Zhang, B. (2019) The 20 biggest airlines in the world, ranked. *Business Insider*, 6 March. Available at: www.businessinsider.com/biggest-airlines-world-oag-2019-3?r=US&IR=T (accessed 6 November 2020).

Index

Note: The page numbers in italics and bold represents figures and tables respectively.